AGRICULTURE TO ZOOLOGY

CHANDOS

INFORMATION PROFESSIONAL SERIES

Series Editor: Ruth Rikowski
(email: Rikowskigr@aol.com)

Chandos' new series of books is aimed at the busy information professional. They have been specially commissioned to provide the reader with an authoritative view of current thinking. They are designed to provide easy-to-read and (most importantly) practical coverage of topics that are of interest to librarians and other information professionals. If you would like a full listing of current and forthcoming titles, please visit www.chandospublishing.com.

New authors: We are always pleased to receive ideas for new titles; if you would like to write a book for Chandos, please contact Dr. Glyn Jones on g.jones.2@elsevier.com or telephone +44 (0) 1865 843000.

AGRICULTURE TO ZOOLOGY
Information Literacy in the Life Sciences

Edited by

JODEE L. KUDEN
JULIANNA E. BRAUND-ALLEN
DARIA O. CARLE

Chandos Publishing is an imprint of Elsevier
50 Hampshire Street, 5th Floor, Cambridge, MA 02139, United States
The Boulevard, Langford Lane, Kidlington, OX5 1GB, United Kingdom

Copyright © 2017 Daria O. Carle, Julianna E. Braund-Allen and Jodee L. Kuden.
Published by Elsevier Limited. All rights reserved

No part of this publication may be reproduced or transmitted in any form or by any means, electronic or mechanical, including photocopying, recording, or any information storage and retrieval system, without permission in writing from the publisher. Details on how to seek permission, further information about the Publisher's permissions policies and our arrangements with organizations such as the Copyright Clearance Center and the Copyright Licensing Agency, can be found at our website: www.elsevier.com/permissions.

This book and the individual contributions contained in it are protected under copyright by the Publisher (other than as may be noted herein).

Notices
Knowledge and best practice in this field are constantly changing. As new research and experience broaden our understanding, changes in research methods, professional practices, or medical treatment may become necessary.

Practitioners and researchers must always rely on their own experience and knowledge in evaluating and using any information, methods, compounds, or experiments described herein. In using such information or methods they should be mindful of their own safety and the safety of others, including parties for whom they have a professional responsibility.

To the fullest extent of the law, neither the Publisher nor the authors, contributors, or editors, assume any liability for any injury and/or damage to persons or property as a matter of products liability, negligence or otherwise, or from any use or operation of any methods, products, instructions, or ideas contained in the material herein.

Library of Congress Cataloging-in-Publication Data
A catalog record for this book is available from the Library of Congress

British Library Cataloguing-in-Publication Data
A catalogue record for this book is available from the British Library

ISBN: 978-0-08-100664-1

For information on all Chandos Publishing publications
visit our website at https://www.elsevier.com/books-and-journals

 Working together to grow libraries in developing countries

www.elsevier.com • www.bookaid.org

Publisher: Glyn Jones
Acquisition Editor: Glyn Jones
Editorial Project Manager: Lindsay Lawrence
Production Project Manager: Surya Narayanan Jayachandran
Designer: Victoria Pearson

Typeset by TNQ Books and Journals

CONTENTS

About the Authors and Editors — *ix*
Preface — *xv*
Acknowledgments — *xix*

1. Introduction to Information Literacy in the Life Sciences — 1
Jodee L. Kuden and Julianna E. Braund-Allen

Defining Information Literacy — 1
Importance of Information Literacy — 2
Information Literacy and Higher Education — 9
Information Literacy and the Sciences — 10
Information Literacy and the Life Sciences — 13
References — 14

2. Scientific Literacy — 17
Elizabeth A. Berman and Jodee L. Kuden

Introduction — 17
Defining Scientific Literacy — 19
Scientific Literacy and Information Literacy — 21
Improving Scientific Literacy Using Information Literacy Skills — 24
References — 25

3. Designing Information Literacy Instruction for the Life Sciences — 27
Katherine O'Clair

Introduction — 27
The Life Sciences Curriculum — 27
Characteristics of Students in the Life Sciences — 30
Collaborating With Instructors — 31
Key Opportunities for Integrating Information Literacy — 31
Information Literacy Competencies — 33
Designing With the Objective–Activity–Assessment Approach — 36
Recommendations and Practical Advice — 42
Conclusion — 43

Recommended Resources	44
Acknowledgments	44
References	44

4. Agriculture and Plant Sciences Information Literacy — 47
Livia M. Olsen

Introduction	47
Literature Review	48
Discipline Resources	50
Information-Seeking Behavior	54
Discipline Information Literacy Instruction	56
Information Literacy Instruction for Graduate Students	58
Emerging Trends and Staying Up-To-Date	58
Conclusion	59
Recommended Resources	60
References	61

5. Marine and Aquatic Sciences Information Literacy — 63
Sally Taylor

Introduction	63
Unique Aspects of Searching	64
Discipline Resources	66
Discipline Information Literacy Instruction	80
Staying Up-To-Date	83
Final Thoughts	84
References	85

6. Polar (Arctic and Antarctic) Sciences Information Literacy — 87
Sandy Campbell, Jessica Thorlakson and Julianna E. Braund-Allen

Why a Chapter on Polar Sciences Information Literacy?	87
Defining Information Literacy in the Polar Context	87
Where to Look for Polar Information	90
A Listing of Polar Information Resources	92
Discipline Information Literacy Instruction	102
Emerging Trends and Staying Up-To-Date	105
Conclusion	107
References	107

7. Zoology and Animal Sciences Information Literacy   **109**
Daria O. Carle

Introduction   109
Taxonomy/Scientific Nomenclature   109
Discipline Resources   110
Discipline Information Literacy Instruction   117
Further Thoughts   123
Emerging Trends and Staying Up-To-Date   125
Conclusion   125
References   126

Appendix 1: Additional Resources   *127*
Index   *135*

ABOUT THE AUTHORS AND EDITORS

Elizabeth A. Berman is the assistant director and head of Research and Instruction at Tisch Library, Tufts University. Her scholarship focuses on the intersection of information literacy, scientific inquiry, and scholarly communications. She was a member of the Association for College and Research Libraries (ACRL) Task Force that developed the ACRL Framework for Information Literacy for Higher Education and has frequently presented on and written about information literacy, in particular through the lens of the sciences.

Elizabeth A. Berman
Assistant Director, Tisch Library
Head of Research and Instruction and Digital Design Lab
Tufts University
35 Professors Row
Medford, MA 02155
elizabeth.berman@tufts.edu

Julianna E. Braund-Allen is an instruction and reference librarian and library liaison for Arctic Studies at the University of Alaska Anchorage (UAA) and a management team librarian at the Alaska Resources Library and Information Services (ARLIS). Professor Braund-Allen is a third-generation Alaskan, a former teamster at Prudhoe Bay on Alaska's North Slope, and has worked with Arctic information in libraries and research settings since 1988, when she began her career at UAA's Arctic Environmental Information and Data Center. In the 1990s, she helped conceive of and establish ARLIS, which brought together the collections, budgets, and library expertise of nine Anchorage-based natural and cultural resources collections from federal, state, and university entities. She continues to co-manage ARLIS, which garnered a National Performance Review Hammer Award and the National Award for Museum and Library Service, among others. Professor Braund-Allen has worked as a reference and instruction librarian at UAA since the early 1990s, where she has focused on the library's one-credit, web-based course, LS101, Introduction to Academic Research. She is senior editor of the *Polar Libraries Bulletin* and an ex-officio member of the Polar Libraries Colloquy.

Julianna E. Braund-Allen
Reference Librarian, UAA/APU Consortium Library

Management Team Librarian, ARLIS
University of Alaska Anchorage
3211 Providence Drive
Anchorage, AK 99508
jebraundallen@alaska.edu

Sandy Campbell is the health sciences librarian at the John W. Scott Health Sciences Library, University of Alberta, where she provides liaison services to the School of Public Health and the Faculty of Medicine and Dentistry. She is the former acting collections manager for the Science and Technology Library at the University of Alberta. She is active in the Polar Libraries Colloquy, serving on its Steering Committee and representing it as a member of the University of the Arctic Council, where she also has chaired the Digital Library Committee. Her particular interests are in alpine and polar information and information literacy. She has presented and published on these and other topics nationally and internationally and is a regular reviewer of Arctic children's books. She is currently examining the role of librarians in systematic review searching and the Digital Library North project.

Sandy Campbell
Health Sciences Librarian
John W. Scott Health Sciences Library
2K3.26 Walter C. Mackenzie Health Science Centre
University of Alberta
Edmonton, AB, Canada T6G 2R7
sandy.campbell@ualberta.ca

Daria O. Carle is a science librarian and library liaison for Engineering, Mathematics, and Natural Sciences at the University of Alaska Anchorage (UAA). She has significant experience in providing library instruction in the sciences and has published peer-reviewed articles in science education journals on incorporating the library into scientific writing courses. Prior to her career in libraries, Professor Carle worked as a seasonal wildlife biotechnician for the federal government. After receiving her MLIS at the University of Wisconsin–Madison, she was selected for the National Library of Medicine's Associate Program, a highly competitive postgraduate internship, and worked at the University of Minnesota's Bio-Medical Library and the University of Colorado at Boulder Science Library. Since her arrival at UAA in 2000, she has successfully built a course-integrated library instruction program in the sciences from the ground up, incorporating elements of information literacy into the classes she teaches. Professor Carle is an editor of the *Polar Libraries Bulletin* and ex-officio member of the Polar Libraries Colloquy, and is also active in the International Association of Aquatic and Marine Science Libraries and Information Centers (IAMSLIC).

Daria O. Carle
Science Librarian
Library Liaison for Engineering, Mathematics, and Natural Sciences
UAA/APU Consortium Library
University of Alaska Anchorage
3211 Providence Drive
Anchorage, AK 99508
docarle@alaska.edu

Jodee L. Kuden is head of Collection Development at the University of Alaska Anchorage (UAA). Professor Kuden has worked in a variety of libraries, with more than 27 years of professional experience at three universities, giving her a diverse background in academic settings. Prior to her MLS, she was a school librarian, a director of a small public library, and worked a few brief stints as a solo librarian in special libraries. Over the years, Professor Kuden has had many responsibilities in university library settings, beginning as a reference and instruction librarian; serving as library liaison to agriculture, extension service, economics, and business; and working part-time in government documents, interlibrary loan, special collections and archives, and indexing departments. For the last nine years as head of Collection Development at UAA, she manages all formats of the collection especially electronic resources, directs the library liaisons collection program and staff, and oversees multimillion dollar library budgets and cooperative arrangements throughout the state. Professor Kuden holds a second master's degree in agricultural education, which aligns with one of her research emphases in agricultural information.

Jodee L. Kuden
Head of Collection Development
UAA/APU Consortium Library
University of Alaska Anchorage
3211 Providence Drive
Anchorage, AK 99508
jlkuden@alaska.edu

Katherine O'Clair is the agriculture and environmental sciences librarian and college librarian for the College of Agriculture, Food and Environmental Sciences at California Polytechnic State University in San Luis Obispo. From 2004 to 2009, she served as the life sciences librarian at Arizona State University in Tempe. Her professional interests include integrating information literacy into the curriculum, assessment of student learning, diversity and inclusion in libraries, and early career mentoring in librarianship. She earned her BS in Environmental Science from Nazareth College in Rochester, New York, and her MS in Library and Information Studies from Florida State University.

> Katherine O'Clair
> Agriculture and Environmental Sciences Librarian
> Robert E. Kennedy Library
> One Grand Avenue, Building 35, Room 216J
> California Polytechnic State University
> San Luis Obispo, CA 93407
> koclair@calpoly.edu

Livia M. Olsen is an academic services librarian and sciences team leader in K-State Libraries' Academic Services Department. She serves faculty and students in the biological sciences and agriculture along with working on a variety of digital projects. She serves as the 2016–17 Chair of the AgNIC Coordinating Committee and K-State Libraries representative to The Rangelands Partnership. She earned her MLS from the University of North Texas and BS in Horticulture from the University of Nebraska–Lincoln.

> Livia M. Olsen
> Academic Services Librarian—Agriculture
> Kansas State Libraries' Academic Services Department Sciences Team Leader
> 126 Hale Library
> Kansas State University
> Manhattan, KS 66506
> livia@ksu.edu

Sally Taylor is a science librarian at the Woodward Library at the University of British Columbia (UBC) in Vancouver, Canada. In addition to providing liaison support and instruction for students and faculty in the biological sciences, fisheries, and forestry, she is a member of a small team that manages the science collections. She received an MLIS from UBC and a BSc Honours and MSc in Biology from Queen's University in Kingston, Ontario. She is a past president of the International Association of Aquatic and Marine Science Libraries and Information Centers (IAMSLIC) and teaches Science and Technology Information Sources and Services at UBC's iSchool.

Sally Taylor
Liaison Librarian for Forestry, Institute for the Oceans and Fisheries, Institute for Resources, Environment and Sustainability, and graduate programs in Botany, Microbiology and Immunology, Zoology
Woodward Library
2198 Health Sciences Mall
University of British Columbia
Vancouver, BC, Canada V6T 1Z3
sally.taylor@ubc.ca

Jessica Thorlakson is a librarian at the Cameron Science and Technology Library at the University of Alberta, providing library subject liaison services to the Faculty of Agriculture, Life, and Environmental Sciences (ALES). Her research interests include digital scholarship (3D printing and data visualization), scholarly communications (research metrics and scholarly identification), and information literacy in academic libraries. She received a BA in English Literature from MacEwan University in 2011 and an MLIS from the University of Alberta's School of Library and Information Studies in 2013.

Jessica Thorlakson
Agriculture, Life, and Environmental Sciences Librarian
Cameron Library, 1-55
116 Street & 85 Avenue
University of Alberta
Edmonton, AB, Canada T6G 2R3
jthorlak@ualberta.ca

PREFACE

Agriculture to Zoology: Information Literacy in the Life Sciences sets the stage for purposefully integrating information literacy activities within the subject-specific content of the life sciences. The book is written for librarians and other professionals who teach information literacy skills, especially those in the science disciplines, and most especially the life sciences. It is also intended to be helpful to secondary school teachers, college faculty who teach life science-related subjects, library school students, and others interested in information literacy and science education. Anyone wanting to learn more about the Earth's life sciences, from citizen to scientist, will benefit as well.

The book's seven chapters fill a gap with varying perspectives of literacy instruction in the life sciences and include resources identified by academic librarians as important for use in subject-specific research in higher education. Contributors are longtime specialists in the fields of the life sciences, science and information literacy, scientific and electronic communication, assessment, and more, including Arctic and Antarctic information.

The Introduction, by Kuden and Braund-Allen, lays out the purpose of the book in the context of information literacy generally, providing a brief discussion of its definition, history, and move toward metaliteracy, as well as a quick look at the international commitment to information literacy. It describes how information literacy supports scientific literacy, accreditation, and lifelong learning. In addition, the chapter illustrates the role that information literacy and collaboration between librarians and discipline faculty can have in higher education, especially in the science disciplines.

The next chapter on Scientific Literacy by Berman and Kuden describes scientific literacy, defines what it is, and discusses how it is used and interpreted by both scientists and the public. The chapter goes on to explain how scientific literacy converges with information literacy and how each contributes to the understanding of scientific information. It describes how scientific literacy is important to interdisciplinary and transdisciplinary research in addition to the science research disciplines.

In Designing Information Literacy Instruction for the Life Sciences, O'Clair further refines the idea of scientific literacy as it relates to teaching information literacy. She describes how information literacy for the life sciences differs from other disciplines and discusses important aspects to consider when planning and developing instruction for these students. The chapter briefly places current information literacy competencies in context and provides recommended resources and practical suggestions for librarians, including opportunities to collaborate with discipline-specific instructors and a carefully thought-out Objective–Activity–Assignment Approach for student learning.

The next several chapters examine information literacy within specific life science subjects, including Agriculture and Plant Sciences (Olsen), Marine and Aquatic Sciences (Taylor), Polar Sciences (Campbell, Thorlakson, and Braund-Allen), and Zoology and Animal Sciences (Carle). Each chapter includes a discussion of the specific life science discipline and an overview of the most important resources used to find information within that discipline including traditional and nontraditional sources, and offers practical examples of activities to teach students the information literacy skills that are essential to their understanding of the discipline.

Olsen discusses in detail the discipline of agriculture and plant sciences and the wide variety of information resources involved when conducting research. The chapter compiles the most important agriculture information sources from publicly available to propriety databases, and it describes the information-seeking behaviors of those involved in agricultural research. It also gives a number of ideas for information literacy instruction for students in the agricultural and plant sciences. The author briefly touches on the aspect of providing information instruction and resources for the agricultural practitioner.

The next chapter by Taylor features marine and aquatic sciences and the often multidisciplinary nature of the research conducted worldwide by academia, industry, and government and nongovernment organizations. The chapter examines unique aspects such as geography to keep in mind when searching for information and emphasizes the importance of publication types beyond journal literature. It highlights key information resources, especially some less common types such as intergovernmental resources and Citizen Science Open Access sites. There are practical examples for teaching information literacy skills to undergraduate and graduate students, as well as advice for staying current in the field.

Another unique chapter concerns polar information, by Campbell, Thorlakson, and Braund-Allen. Not many sources include information on researching polar topics, specifically identifying resources for finding Arctic and Antarctic information. The authors describe how locating such information is inherently difficult because of the way that traditional (i.e., western) organizational systems group information, essentially by subject rather than geography. The chapter introduces techniques and methodologies for identifying polar resources within traditional information systems, and it identifies resources that focus specifically on polar information. It also highlights traditional knowledge as an important source of information in polar research.

Carle describes the discipline of zoology and animal sciences and the many subdisciplines it comprises. The chapter identifies core databases in the zoological discipline and gives a selected listing of other resources that include information related to the topic. Techniques for teaching information literacy skills to undergraduate and graduate zoology students, along with sources for finding additional material on information literacy in the life sciences, are also given.

The book concludes with a short list of additional resources compiled by Kuden. It includes information resources relevant to scientific information but not mentioned in the previous chapters, and sources of alternative literature, where students might find information from a different point of view or to support an ethical or philosophical stance, or other material not found in the mainstream media or traditional publications.

It is our sincere hope that this book will provide resources, tools, and ideas that anyone can use in pursuit of improving information literacy instruction, especially in the life sciences. It is intended to serve as a guide for those wishing to enhance their knowledge and instruction of information literacy as well as an entry point into the instruction for those new to the field. May we all continue to be lifelong learners in our quest to become better instructors.

Jodee L. Kuden, Julianna E. Braund-Allen, and Daria O. Carle

ACKNOWLEDGMENTS

Many thanks to our contributing authors—Elizabeth A. Berman, Katherine O'Clair, Livia M. Olsen, Sally Taylor, Sandy Campbell, and Jessica Thorlakson—all of whom are not only practicing librarians and experts in their fields but have also made the time in their busy schedules to generously share their knowledge and best practices with us in the chapters contained in this book. We thank our outside reader, Dr. Nancy F. Carter, retired mathematics/physics librarian, University of Colorado at Boulder, for her helpful insights. We are grateful to the support of the University of Alaska Anchorage, especially our UAA/APU Consortium Library Dean, Steve Rollins. Thanks also to Lindsay Lawrence, our editorial project manager, and her team at Chandos Imprint, Elsevier. We especially appreciate their patience in working with us and their belief in the importance and timeliness of the topic of information literacy in the life sciences.

CHAPTER 1

Introduction to Information Literacy in the Life Sciences

Jodee L. Kuden, Julianna E. Braund-Allen
University of Alaska Anchorage

Strong information literacy skills equip students with the potential to grow into capable individuals and scientists better able to meet future challenges. *Agriculture to Zoology* provides direction for librarians to incorporate information literacy into life sciences courses and programs that will give students opportunities to demonstrate learning outcomes in both the life sciences and information literacy.

    The editors and authors are all librarians with knowledge and experience working in the life sciences and collaborating with discipline faculty. Their goal is to make this book informative and useful in a practical sense, with each chapter describing a hands-on approach to information literacy for this specialized area of study. Since the life sciences cover so many different topics, only a few could be addressed directly, but much of what is written here can be adapted to other areas. The chapters identify the main resources and provide examples of successful strategies and learning activities. They offer explanations and ways for assessing success of teaching information literacy and confirming that outcomes are met. This book is not an exhaustive resource on teaching information literacy to life sciences students; however, it does provide pointers to tools, resources, and an array of possible approaches to information literacy instruction.

DEFINING INFORMATION LITERACY

As people around the globe tap into the Web, the need to understand the information they have discovered creates learning opportunities. The increased amount and availability of information plays directly into the concept of "information literacy," defined by the American Library Association (ALA) in 1989: "a person must be able to recognize when information is needed and have the ability to locate, evaluate, and use effectively the needed information." A decade later, the Association of College and Research

Libraries (ACRL), a division of ALA, developed a set of information literacy standards for higher education based on this definition (2000). These standards were a defined set of competencies with performance indicators and outcomes.

More recently, ACRL has updated and expanded this definition in their 2016 Framework.

> Because this Framework envisions information literacy as extending the arc of learning throughout students' academic careers and as converging with other academic and social learning goals, an expanded definition of information literacy is offered here to emphasize dynamism, flexibility, individual growth, and community learning: Information literacy is the set of integrated abilities encompassing the reflective discovery of information, the understanding of how information is produced and valued, and the use of information in creating new knowledge and participating ethically in communities of learning.
> **ACRL Board (2016)**

The definition can be applied to any type of information—oral, print, or digital—and it can be applied to people with or without Internet access. All sectors of the library profession, from academic and school to public and special, have been presenting and writing about information literacy for decades. International associations and agencies, federal governments, higher education institutions, and other groups from around the world have embraced the information literacy definitions from ALA and ACRL.

IMPORTANCE OF INFORMATION LITERACY
History

The term *Information Age* is frequently used in describing the late decades of the 20th century and the 21st century to date. James R. Messenger, known as the "Father of the Information Age," was the first to write about this phenomenon in 1982:

> The Information Age is a true new age based upon the interconnection of computers via telecommunications, with these information systems operating on both a real-time and as-needed basis. Furthermore, the primary factors driving the new age forward are convenience and user-friendliness, which, in turn, will create user dependence. User dependence is what will ensure the full implementation of the technological platform that will become the foundation for a new economy, and dependence upon information systems is what will eventually distinguish the Information Age from the Industrial Age in the same manner that reliance upon mass production manufacturing techniques distinguished the Industrial Age from Agrarian Society (p. 1).

Just as transportation in the Aviation Age transformed the first half of the 20th century, so has the Information Age rapidly altered this current century, and it continues to accelerate. In the early 20th century, airplanes opened up access to the world, making it physically possible for people to reach any corner of it, no matter how remote. In this Information Age, the World Wide Web (WWW or the Web), the Internet, telecommunication systems, and huge amounts of data have opened even the furthermost regions of the world to information, making virtual communication across the globe not only possible, but also often instantaneous. This digital revolution, the movement behind the Information Age, continues to create new industries, job sectors, equipment, and e-commerce.

Now nearly 20 years into the 21st century, a pressing goal for many countries and international organizations has been to expand mobile telecommunication systems to reach more people in more areas. As one example, the International Telecommunications Union (ITU) of the United Nations "is committed to connecting all the world's people—wherever they live and whatever their means. Through our work, we protect and support everyone's fundamental right to communicate" (2016, Paragraph 2). Since ITU began this work in 1992, millions more have gained Internet access, continuing the Information Age transformation.

Nevertheless, one must recognize that the availability to such information is limited to those who can afford Internet access, just as being able to ride in an airplane has always been defined by affordability. When people have access to the Web, the amount of information available to them explodes. While this can result in what is commonly known as information overload and render an individual unable to discern which pieces of information matter to them, it also presents individuals and groups with untold possibilities for discovering, learning, and understanding a plethora of topics—as well as opportunities for synthesizing and disseminating new information and knowledge. The availability of information has tremendous influence and powerfully affects everything we do.

Building the Case for Information Literacy

The importance of information literacy, not just for students but across all sections of society, has been a central theme of the library profession for much longer than the last 30 years; librarians have been involved in preserving, providing access, and teaching others how to find and best use information for ages. It is no wonder that in this particular age of information transformation, librarians have been in the vanguard of researching, presenting, and writing about the importance of information literacy. During this time, the concept of information literacy has gradually been recognized by professional societies and associations outside of librarianship. Other fields may not necessarily state the concept directly as information literacy; more often, professional organizations use terms or make statements that refer to literature skills or information skills in their field or body of knowledge. Some broader terms such as lifelong learning or workforce skills have long been accepted in fields such as human resources, and they are now often applied more broadly beyond specific professional societies or disciplines.

Information Literacy Supports Lifelong Learning

Information literacy as a term was coined by P.G. Zurkowski in 1974 when he was president of the Information Industry Association (National Forum on Information Literacy, 2016); however, the general population is more familiar with *lifelong learning*, and the concept behind this term has been around far longer than that of information literacy. The *English Oxford Living Dictionaries* define lifelong learning as "a form of or approach to education which promotes the continuation of learning throughout adult life, esp. by making educational material and instruction available through libraries, colleges, or information technology" (Lifelong Learning, 2017). By this definition, information literacy is a cornerstone of lifelong learning.

Project Information Literacy, a 2-year large-scale study conducted by Dr. A. Head (2014), interviewed recent college graduates about lifelong learning. For the purpose of her study, she defined lifelong learning as the kind of purposeful and ongoing learning that has the aim of improving skills, knowledge, or competencies in three areas for (1) use in the workplace, (2) engagement in civic and community activities, and (3) participation in activities for social and personal enrichment. This study reinforces the closeness of information literacy to the concept of lifelong learning. In the Phase One study report, one of the key findings identified the skill of information literacy as necessary to achieve lifelong learning objectives or outcomes:

> **5. Adaptable Information Practices from College**: *Many of the recent graduates we interviewed credited college with "teaching them to learn how to learn,"*

while giving them the confidence to "learn anything on their own." They referred to the critical thinking skills they had taken from college, especially the ability to sort through large volumes of content and synthesize key points, determine bias on web sites and in news articles, evaluate the authority and credibility of sources, and to be flexible and revise search strategies as new information is presented. At the same time, some interviewees did not see how their major had taught them skills that matched their real world needs. Sometimes this was because of differences in their major compared to career, but also because the goals of a college course can be very different from those of the workplace and life away from the campus. Evaluating the usefulness of their college education, a number of interviewees said it was capstone projects, a senior thesis, or involvement in extracurricular activities that nurtured their initiative, curiosity, independent learning, motivation, and involvement. After graduation, they had come to realize these learning dispositions were transferable and critical to their success both as employees and as lifelong learners (p. 5).

Many professional societies recognize the need for lifelong learning skills for their members and offer courses, seminars, conferences, or other learning opportunities. Training individuals in the development of skills that support careers and other needs during their life span is a primary goal of information literacy. An information literate individual can review a large amount of information quickly, honing in on the relevant materials and thus avoiding the information overload that many people experience when searching the Web. When a person recognizes the need for information, can articulate it, and apply information literacy concepts to it, they advance their lifelong learning skills. No matter whether one receives formal training or develops these skills through self-learning, information literacy is needed to fulfill a commitment to lifelong learning in this Information Age.

Information Literacy Supports Accreditation Requirements

Another area into which information literacy has found its way involves the requirements of the six regional educational accreditation bodies for the schools, colleges, and universities in the United States. The two examples following do not use the exact phrase; however, the specific standards allude to exactly the same concept found in ALA's definition of information literacy. In its Criteria for Accreditation, the North Central Association of Schools and Colleges, Higher Learning Commission, states: "Every degree program offered by the institution engages students in collecting, analyzing, and communicating information; in mastering modes of inquiry or creative work; and in developing skills adaptable to changing environments" (2013;

Number CRRT.B.10.010, Section 3.B.3). In its accreditation requirements, the Northwest Commission on Colleges and Universities states:

> Consistent with its mission and core themes, the institution provides appropriate instruction and support for students, faculty, staff, administrators, and others (as appropriate) to enhance their efficiency and effectiveness in obtaining, evaluating, and using library and information resources that support its programs and services, wherever offered and however delivered.
>
> **Northwest Commission on Colleges and Universities (2010), Section 2.E.3**

Many accreditation bodies, be they regional or professional, have embraced concepts of information literacy in various ways. They know that in this Information Age people need a set of skills and abilities to consume information through a critical synthesis to formulate knowledge and communicate effectively. In the United States, librarians are meeting this challenge through information literacy programs found in schools, universities, and public and special libraries. Educational institutions are slowly but steadily adopting information literacy concepts across the curriculum, and they are increasingly pairing classroom teachers or discipline faculty with librarians to do so. Information literacy is crucial to the end goal for students to be successful in their fields of study and careers, and as lifelong learners and citizens of a democracy.

International Support for Information Literacy

The International Federation of Library Associations (IFLA), an influential organization of the information profession, has a longstanding history of discussing and defining information literacy. The IFLA Information Literacy Section developed its own set of standards and mapped these standards to learning objectives. The Section also hosts a blog, intermittently holds conferences or webinars on related topics, and publishes articles and books on media and information literacy. These can be found on its site, https://www.ifla.org/information-literacy.

Throughout Europe and many other regions of the world, librarians and educators have embraced the model of information literacy and created processes to implement it into the education system. Information literacy has been embedded mainly in local education systems, but over the last decade, the model and processes gradually have been included in governmental education requirements. The cooperative International Association for the Evaluation of Educational Achievement (IEA) has made efforts to improve worldwide education since the late 1950s. Over time, it has conducted large cross-national studies and published a variety of documents providing study results, learning strategies, outcomes, assessment, and competencies for students.

One such study, designed to shed light on how well-prepared students are for the Digital Age, is the ICILS: International Computer and Information Literacy Study, which performed a pioneering, comparative assessment of how well young students were progressing toward necessary competencies regarding digital information and products through a survey of their computer and information literacy skills. The study was conducted in various countries between 2010 and 2013. According to Fraillon, Ainley, Schulz, Friedman, and Gebhardt (2014), who published a final report of that assessment cycle, the study gathered data from 60,000 eighth grade students in approximately 3300 schools as well as data from 35,000 teachers and others in those schools. In an infographic published in 2013, IEA highlighted a snapshot point from the ICILS: "most students think they are experienced ICT [information and communication technology] users." It went on to state how the students characterized themselves: "36% have been using computers since they were 6.5 years of age or younger, 89% feel confident to find information on the internet, and 2% [are engaging in] critical thinking while searching for information on line" (IEA, 2013). Considering the age of these students (13–14 years old) at the time of the assessment, there is an excellent educational opportunity to expand their critical thinking skills using information literacy concepts throughout the remainder of their schooling years. Revision of the ICILS began in 2015, and it has an anticipated resurvey date of 2018 to include new schools or school districts and to allow the 2013 participating schools to compare data and determine changes in their students over time. Details about this and additional IEA studies, as well as the 2017 launch of its Open Access gateway and other of its freely accessible materials, can be found on its site, http://www.iea.nl/research-collaboration.

In Moscow in 2012, the United Nations Educational, Scientific, and Cultural Organization (UNESCO) held an International Conference on Media and Information Literacy for Knowledge Societies that resulted in the Moscow Declaration on Media and Information Literacy. This work expanded the ALA definition (1989). One of the points of agreement by participants was that media and information literacy (MIL):

> is defined as a combination of knowledge, attitudes, skills, and practices required to access, analyze, evaluate, use, produce, and communicate information and knowledge in creative, legal and ethical ways that respect human rights. Media and information literate individuals can use diverse media, information sources and channels in their private, professional and public lives. They know when and what information they need and what for, and where and how to obtain it. They understand who has created that information and why, as well as the roles, responsibilities and functions of media, information providers and memory institutions. They

> *can analyze information, messages, beliefs and values conveyed through the media and any kind of content producers, and can validate information they have found and produced against a range of generic, personal and context-based criteria. MIL competencies thus extend beyond information and communication technologies to encompass learning, critical thinking and interpretive skills across and beyond professional, educational and societal boundaries. MIL addresses all types of media (oral, print, analogue and digital) and all forms and formats of resources.*
>
> **UNESCO (2012; Point 2)**

The Moscow Declaration references building its definition on important earlier international documents and agreements, including the 2003 Prague Declaration, Towards an Information Literate Society; the 2005 Alexandria Proclamation, Beacons of the Information Society; the 2011 Fez Declaration on Media and Information Literacy; and the 2011 IFLA Media and Information Literacy recommendations. It goes on to state that "MIL underpins essential competencies needed to work effectively towards achievement of the UN [United Nations] Millennium Development goals, the UN Declaration on Human Rights, and the goals promoted by the World Summit on the Information Society" (UNESCO, 2012; Point 3).

When UNESCO leads an educational change, many countries will then take steps to incorporate UNESCO's goals and concepts into their own educational systems. Results of a Web search for information literacy models, standards, and goals demonstrate and display evidence that countries are adopting the changes. In November 2016, Brazil hosted the fifth annual Global MIL Conference sponsored by UNESCO. Right on the heels of that conference, in December 2016, UNESCO hosted the International Conference of Non-Governmental Organizations, whose theme was NGOs and the Digital Revolution. The challenges discussed focused on access to the Web, e-learning opportunities, and ethical implications of science in the digital revolution. One notable point from the Conference worth mentioning is that "NGOs emphasized the importance of access to reliable information, countering the unprecedented emergence of fake news in the last few months, a field where NGOs could play a monitoring role" (UNESCO, 2016).

Advances in telecommunications, Internet access, and technologies for mobile devices have given the public far greater access to information worldwide than ever before, including Massive Open Online Courses (MOOCs), TED Talks, and YouTube tutorials. For individuals, whether managing their personal or professional lives, this access can provide lifelong learning opportunities to self-direct their learning and awareness of current events, their development of skills and abilities, and their career advancement.

For students, building information literacy skills throughout their years of education gives them the capabilities needed to fulfill their potential to be successful employees, employers, and productive citizens.

INFORMATION LITERACY AND HIGHER EDUCATION

In practical terms, information literacy instruction gives higher education students the currency to go beyond general Google searches and social media. It teaches them to think about information as an entity and pose the "how, what, where, why, who, and when" kinds of questions important to finding information, and filtering and evaluating it to reap something manageable, usable, and authoritative. The 2016 ACRL Framework gives librarians a structure for developing a curriculum program and learning outcomes to advance students' knowledge and skills in today's informational context and forward. The Framework is refined and multidimensional, and yet it allows for leeway.

The six ACRL frames (for example, Authority is Constructed and Contextual, Research as Inquiry, and Searching as Strategic Exploration), along with their associated knowledge practices and dispositions, provide a useful structure for engaging students in critical thinking skills. At the same time, the frames provide a means for grounding students in the basics of information literacy, and beyond to more sophisticated understanding and practices. The concept of metaliteracy running throughout the ACRL Framework is reflected in the nature of the databases and repositories that hold an ever-growing proportion of scientific research. These digital structures, many of which are presented in the ensuing chapters, are often collaborative, multidimensional, and participatory. They allow richer probing for and sometimes transformative uses of the data and research as new technologies become available and are applied. As students interact with the databases and repositories, they engage with this multidimensional and participatory collaboration; they learn from it as they reflect on their search strategies or become participants in citizen science.

As society moves into the Post–Information Age, instruction toward metaliteracy is becoming the next challenge. Metaliteracy is an expansion of information literacy to include the comprehensive set of skills and literacies (visual, media, and so on) needed to effectively engage in shared, interactive digital technologies and collective spaces. Embracing transformative access to information through mobile devices; understanding new collaborative digital communication venues such as open forums, eg., iCitizen and social media; and learning to manage this informational deluge are the basic foundations of metaliteracy. The accomplished metaliterate student would incorporate these modalities for consuming, producing, and sharing information into usable and valuable knowledge (Mackey & Jacobson, 2014, 2017).

As knowledge has expanded and collaboration has grown among disciplines, development of interdisciplinary and transdisciplinary research also reinforces the need for students to learn more advanced information literacy skill sets to engage with this complex literature. One common definition of interdisciplinary research is that it is a study that uses or borrows design or methodology from one discipline that is incorporated into another discipline, requiring the collaboration of researchers from those disciplines, with results interpreted within those disciplinary frameworks. Transdisciplinary research expands interdisciplinary boundaries on all levels and at all stages of the research process, not only using already developed methodology but also creating cross-cutting designs or methodologies. Moreover, the group studying the research question is increased beyond the discipline researchers to include stakeholders; these are community members or other individuals who could be affected by the results. Another aspect is that the collaboration of the researchers expands interpretation and translation beyond their specific disciplines, creating new perspectives, ideas, and even new disciplines. Students may participate in these research projects as researchers or as stakeholder members. In either capacity, students would need to draw on their skill sets and abilities learned through grounding in information literacy and critical thinking.

INFORMATION LITERACY AND THE SCIENCES
Sciences in General

The general concepts of information literacy work across all of the subjects and disciplines found in an educational setting. Just over a decade ago, ACRL's Science and Technology Section (STS) Task Force laid the groundwork for scientific information literacy when it developed standards specific to the sciences and engineering/technology disciplines. The definition provides an excellent framework for narrowing the earlier lens of information literacy to the sciences:

> *Information literacy in science, engineering, and technology disciplines is defined as a set of abilities to identify the need for information, procure the information, evaluate the information and subsequently revise the strategy for obtaining the information, to use the information and to use it in an ethical and legal manner, and to engage in lifelong learning. Information literacy competency is highly*

> *important for students in science and engineering/technology disciplines who must access a wide variety of information sources and formats that carry the body of knowledge in their fields. These disciplines are rapidly changing and it is vital to the practicing scientist and engineer that they know how to keep up with new developments and new sources of experimental/research data.*
> **ALA/ACRL/STS Task Force on Information Literacy for Science and Technology (2006), Paragraph 1**

The introduction goes on to state:

> *Science, engineering, and technology disciplines require that students demonstrate competency not only in written assignments and research papers but also in unique areas such as experimentation, laboratory research, and mechanical drawing. Our objective is to provide a set of standards that can be used by science and engineering/technology educators, in the context of their institution's mission, to help guide their information literacy-related instruction and to assess student progress (Paragraph 3).*

Recognizing the need for future scientists to be equipped for increasingly complex scientific investigation, the Task Force specified that its standards would need to undergo periodic review and should be considered a living document. Along with performance indicators and specific outcomes, the Task Force included five individual standards: An information literate student can (1) determine the nature and extent of the information needed; (2) acquire needed information effectively and efficiently; (3) critically evaluate the procured information and its sources, and as a result, decide whether or not to modify the initial query and/or seek additional sources and whether to develop a new research process; (4) understand the economic, ethical, legal, and social issues surrounding the use of information and its technologies and either as an individual or as a member of a group, use information effectively, ethically, and legally to accomplish a specific purpose; and (5) understand that information literacy is an ongoing process and an important component of lifelong learning and recognize the need to keep current regarding new developments in his or her field (2006).

Collaboration Between Librarians and Discipline Faculty

Many studies have reported on the importance of collaboration between librarians and subject discipline faculty as an essential element of successful integration of information literacy principles into the science classroom. A study by Ferrer-Vincent and Carello (2011) reports on an undergraduate, discipline-specific library instruction program that was embedded in the first-year general biology laboratory course at their university. A follow-up survey of students in advanced biology courses found that those who had experienced the embedded library instruction performed better by a number of measures 3 years later. Students were more likely to take a scholarly approach in their research, made better use of library databases, and better met teacher expectations.

A study by Fuselier and Nelson (2011) reported similar findings for a single-session information literacy lesson, including writing components, also given in an introductory biology laboratory course. Students in sections that received the instruction performed better at identifying primary and secondary sources and in preparing citations than students who had not. A follow-up study showed that the gains in information literacy skills for those who had been exposed to the single-session instruction persisted even a year later. Thompson and Blankinship report similar findings for a weekly, hour-long required sophomore-level course, Biological Literature, which included an embedded librarian. The improvement in research and analytical skills the students gained over three sessions of information literacy instruction, which included hands-on practice and discussion as well as homework assignments, lasted into their upper-division science courses (2015).

Primary Scientific Literature

For librarians to be effective instructors in general, and especially when collaborating with faculty in the sciences or life sciences, they need to understand concepts of scientific literacy and the connections with information literacy. As discussed in the chapters that follow, a person who is scientifically literate is someone who can seek and evaluate science information and is able to engage in discussion or discourse to gain an understanding and perceive the usefulness of scientific developments and advancements, whether these be beneficial or detrimental to the individual, community, or society. In many science disciplines there is a growing trend to use primary literature as a way to teach students the key concepts of scientific literacy, two of which are communication and critical thinking skills. The latter has been described as "the intellectually disciplined process of actively and skillfully conceptualizing, applying, analyzing, synthesizing, and/or evaluating information gathered from, or generated by, observation, experience, reflection, reasoning, or communication, as a guide to belief and action" (Foundation for Critical Thinking, 2015).

Dr. L. Diener, Sciences Department Chair and Biology Program Director at Mount Mary University in Milwaukee, Wisconsin, wrote "When I think about the skills I want my students to leave college with, the ability to think critically is prominently displayed at the top of the list. Thinking about my own development as a scholar I can see that reading, writing and learning to understand the primary scientific literature are some of the activities that helped me to develop those skills" (2015, Paragraph 1). Critical thinking is foundational to information literacy as well, and teaching these skills through the use of primary literature is an excellent learning goal that (1) is easily accessible to librarians when designing information literacy course components and (2) gives students opportunities to develop and expand their knowledge on scientific topics.

Brownell, Price, and Steinman (2013) state that "primary scientific literature is the gold standard by which scientists communicate their results to other scientists, so exposure to and practice dissecting primary literature are important corollaries to communication skills" (p. 1). Life sciences faculty across the globe are writing curriculum and course content teaching scientific literacy through the use of primary literature (Brickman et al., 2012; Bromme, Scharrer, Stadtler, Hömberg, & Torspecken, 2015; Krontiris-Litowitz, 2013). These articles are prime resources for librarians to use as discussion points with life sciences faculty when planning or incorporating information literacy into their classes. The articles give examples of exercises, assignments, and outcomes that could be revamped to work with local curriculum and courses.

INFORMATION LITERACY AND THE LIFE SCIENCES

The unique aspects of the life sciences can challenge students and librarians alike. As illuminated in the chapters that follow, the complexity of the life sciences—studying living organisms in the six kingdoms of plants, animals, protozoa, fungi, archaebacterial, and eubacteria—encompasses literature in many formats and venues. The literature can be found in well-known or obscure collaborative databases; in vast data sets, interactive maps, and statistical groupings; in globally shared repositories; and in government reports, academic journals, technical reports and bulletins, popular magazines, and archives. The literature includes all formats, even diaries, videos, and microfiche—the latter considered old fashioned by many, but an excellent archival medium, and sometimes containing the only existing copy of a unique resource. The literature also can be difficult to locate because it may be organized geographically or consist of grey literature of little or limited distribution that does not appear in resources searchable in the mainstream databases or through other finding aids.

The chapters delve into this literature of the life sciences and into ways that librarians, alone and in collaboration with discipline faculty, can thread the concepts of information literacy through the tangle of its specialized research. A discussion of the intersection of scientific and information literacy leads to an exposition on practical considerations for designing information literacy instruction that is tailored to the life sciences. The remaining sections flesh out an information literacy primer for the life sciences covering its vast resources; tried-and-true instructional insights, models, exercises, and activities; emerging trends in the field; and ways to stay in touch with current developments. Contributing librarians, whose expertise literally ranges from the Earth's oceans and freshwater environments to the poles, from our planet's agriculture and plant sciences to its zoology and animal sciences, welcome you to this fascinating field.

REFERENCES

ALA/ACRL/STS Task Force on Information Literacy for Science and Technology. (2006). *Information literacy standards for science and engineering/technology*. Retrieved from http://www.ala.org/acrl/standards/infolitscitech.

American Library Association [ALA]. (1989). *Presidential Committee on Information Literacy: Final report*. Retrieved from http://www.ala.org/acrl/publications/whitepapers/presidential.

Association of College and Research Libraries [ACRL]. (2000). *Information literacy competency standards for higher education*. Retrieved from http://www.ala.org/acrl/standards/informationliteracycompetency.

Association of College and Research Libraries [ACRL] Board. (2016). *Framework for information literacy for higher education*. Retrieved from http://www.ala.org/acrl/standards/ilframework.

Brickman, P., Gormally, C., Francom, G., Jardelezz, S. E., Schutte, V. G. W., Jordan, C., et al. (2012). Media-savvy scientific literacy: Developing critical evaluation skills by investigating scientific claims. *American Biology Teacher, 74*(6), 374–379. Retrieved from http://www.jstor.org/stable/10.1525/abt.2012.74.6.4.

Bromme, R., Scharrer, L., Stadtler, M., Hömberg, J., & Torspecken, R. (2015). Is it believable when it's scientific? How scientific discourse style influences laypeople's resolution of conflicts. *Journal of Research in Science Teaching, 52*(1), 36–57. http://dx.doi.org/10.1002/tea.21172.

Brownell, S. E., Price, J. V., & Steinman, L. (2013). A writing-intensive course improves biology undergraduates' perception and confidence of their abilities to read scientific literature and communicate science. *AJP: Advances in Physiology Education, 37*, 70–79. http://dx.doi.org/10.1152/advan.00138.2012.

Diener, L. (2015). *UG Confab: A LifeSciTRC SciEd blog: Using the primary literature in your classroom*. Retrieved from http://blog.lifescitrc.org/ugconfab/2015/11/17/using-the-primary-literature-in-your-classroom/.

Ferrer-Vincent, I. J., & Carello, C. A. (2011). The lasting value of an embedded, first-year, biology library instruction program. *Science & Technology Libraries, 30*(3), 254–266. http://dx.doi.org/10.1080/0194262X.2011.592789.

Foundation for Critical Thinking. (2015). *Defining critical thinking*. Retrieved from http://www.criticalthinking.org/pages/defining-critical-thinking/766.

Fraillon, J., Ainley, J., Schulz, W., Friedman, T., & Gebhardt, E. (2014). *Preparing for life in a digital age: The IEA International Computer and Information Literacy Study International Report* Open Access. Retrieved from http://link.springer.com/book/10.1007%2F978-3-319-14222-7.

Fuselier, L., & Nelson, B. (2011). A test of the efficacy of an information literacy lesson in an introductory biology laboratory course with a strong science-writing component. *Science & Technology Libraries*, 30(1), 58–75. http://dx.doi.org/10.1080/0194262X.2011.547101.

Head, A. J. (2014). *Project information literacy's lifelong learning study, phase one: Interviews with recent graduates, research brief*. Retrieved from http://www.projectinfolit.org/uploads/2/7/5/4/27541717/pil_phase1_trends_lll_7_2014.pdf.

International Association for the Evaluation of Educational Assessment (IEA). (2013). *Students in the digital age: Current state of play, assessing student's computer and information literacy skills infographic*. Retrieved from http://www.iea.nl/fileadmin/user_upload/Studies/ICILS_2013/ICILS_2013_infographic.pdf.

International Telecommunications Union (ITU). (2016). *About ITU*. Retrieved from http://www.itu.int/en/about/Pages/overview.aspx.

Krontiris-Litowitz, J. (2013). Using primary literature to teach science literacy to introductory biology students. *Journal of Microbiology & Biology Education*, 14(1), 66–77. http://dx.doi.org/10.1128/jmbe.v14i1.538.

Lifelong learning. (2017). In *English Oxford living dictionaries*. Oxford, UK: Oxford University Press. Retrieved from https://en.oxforddictionaries.com/definition/lifelong_learning.

Mackey, T. P., & Jacobson, T. E. (2014). *Metaliteracy: Reinventing information literacy to empower learners.* Chicago: Neal-Schuman.

Mackey, T. P., & Jacobson, T. E. (2017). *Metaliteracy*. Retrieved from https://metaliteracy.org.

Messenger, J. R. (1982). *The theory of the information age* Reprinted Excerpt. Retrieved from http://www.informationage.org/.

National Forum on Information Literacy. (2016). *Paul G. Zurkowski blog post*. Retrieved from http://infolit.org/paul-g-zurkowski/.

North Central Association of Schools and Colleges, Higher Education Commission. (2013). *Policy: Criteria for accreditation* Number CRRT.B.10.010. Retrieved from http://policy.hlcommission.org/Policies/criteria-for-accreditation.html.

Northwest Commission on Colleges and Universities. (2010). *Standard two: Resources and capacity* 2.E Library and Information Resources. Retrieved from http://www.nwccu.org/Standards%20and%20Policies/Standard%202/Standard%20Two.htm.

Thompson, L., & Blankinship, L. A. (2015). Teaching information literacy skills to sophomore-level biology majors. *Journal of Microbiology & Biology Education*, 16(1), 29–33. http://dx.doi.org/10.1128/jmbe.v16i1.818.

United Nations Educational, Scientific, and Cultural Organization (UNESCO). (2012). *The Moscow declaration on media and information literacy*. Retrieved from http://www.unesco.org/new/fileadmin/MULTIMEDIA/HQ/CI/CI/pdf/In_Focus/Moscow_Declaration_on_MIL_eng.pdf.

United Nations Educational, Scientific, and Cultural Organization (UNESCO). (2016). *NGOs and the digital revolution* News Release. Retrieved from http://en.unesco.org/news/ngos-and-digital-revolution-0.

CHAPTER 2

Scientific Literacy

Elizabeth A. Berman[1], Jodee L. Kuden[2]
[1]Tufts University, Medford, MA, United States; [2]University of Alaska Anchorage, Anchorage, AK, United States

INTRODUCTION

Traditionally, scholarship falls into one of three branches of learning: the humanities, the social sciences, or the sciences. Science disciplines are further broken down into either basic or pure science (such as physics or chemistry), which describes the phenomena of the material universe and their laws, or applied science (such as engineering), which emphasizes practical applications of scientific discoveries; some fields (such as agriculture) fall somewhere in between. Students in the sciences and engineering are expected to demonstrate competency in areas such as experimentation, laboratory research, simulation, fieldwork, and mechanical drawing, ultimately producing technical reports, scientific papers and presentations, lab reports, data sets, and prototypes.

Yet these disciplines are rapidly changing. Researchers have begun addressing scientific problems using collaborative and cross-disciplinary approaches, such as the emergent fields of bioinformatics and biomathematics, where computer science and mathematical modeling is used to understand biology, medicine, and biotechnology. According to the National Research Council (NRC), "[f]urther research at this intersection not only will advance our understanding of the fundamental questions of science, but will also significantly impact public health, technology, and stewardship of the environment for the benefit of society" (2010, p. 4). The National Science Foundation recognizes the growing arena of interdisciplinary research: It gives high priority to interdisciplinary grant applications and supports specific solicitations for various interdisciplinary research (National Science Foundation, 2016).

Another trend can be seen in the growth and expansion of transdisciplinary research, as exemplified by Harvard's Transdisciplinary Research in Energetics and Cancer Center in Boston, which integrates discipline-specific research, levels of influence, and interventions in pursuit of a cohesive understanding and treatment of cancer. Transdisciplinary research can be considered distinct from interdisciplinary research in that its emphasis is on a holistic approach intended to cut across individual disciplines to reach a higher perspective of knowledge; it is "conducted by investigators from different disciplines working jointly to create new conceptual, theoretical, methodological, and translational innovations that integrate and move beyond discipline-specific approaches to address a common problem" (Harvard Transdisciplinary Research in Energetics and Cancer Center, 2017).

The emergence of e-Science similarly reflects this rapidly changing scholarly landscape. According to the National e-Science Centre (n.d.), e-Science is defined as "the large scale science that will increasingly be carried out through distributed global collaborations enabled by the Internet. Typically, a feature of such collaborative scientific enterprises is that they will require access to very large data collections, very large scale computing resources, and high performance visualization back to the individual user scientists" (Paragraph 1). According to Berman (2013a), e-Science seeks new partnerships and collaborations beyond traditional silos to solve complex problems; it embodies new models of information production; and it requires a diverse set of professional skills and literacies that intersect and interact within an evolving information landscape.

There is also an increasing recognition that students must move beyond scientific and technical expertise in these fields and develop an understanding of the cultural, social, aesthetic, and political aspects of scientific issues (Lupton, 2008; Quinn, Burbach, Matkin, & Flores, 2009). From human health to agriculture to engineering, a framework that integrates the social sciences and humanities with the sciences to actively seek solutions to our most complex problems underpins the necessity of a holistic approach to scientific research. According to Vincent and Focht (2009), the goal of these programs should be to train "interdisciplinarians" by exposing students to a wide range of study in the natural and social sciences, the humanities (including ethics, history, and human cultures), and topics from applied fields (including business).

In addition to an increased focus on interdisciplinary academics, accreditation boards and undergraduate programs in the sciences no longer expect graduates to master only scientific or technical competencies. There is an expectation that students graduate with a suite of professional or "soft" skills, including written and oral communication skills, problem-solving skills, teamwork/conflict management skills, project management skills, and critical and creative thinking skills (ABET Engineering Accreditation Commission, 2015; American Association for the Advancement of Science,

2010; American Chemical Society, 2015; Institute of Physics, 2011; National Research Council, 2009). In a 2015 survey from the Association of American Colleges and Universities, researchers found that employers consider written and oral communication skills, teamwork skills, ethical decision making, critical thinking, and the ability to apply knowledge in real-world settings as essential attributes to long-term career success.

Thus emerges a nuanced picture of scientific education that necessitates students to understand and dissect a basic scientific or technical concept and comprehend the interrelationship of science and society. This concept is known as *scientific literacy*.

DEFINING SCIENTIFIC LITERACY

Scientific literacy, also referred to as *science communication* and *public communication of science*, has been discussed at great length in the scholarly literature since the 1950s, but lacks a singular definition. One broad category of definitions emphasizes the importance of fact-based scientific knowledge and the scientific method—what the public ought to know about science (Burns, O'Connor, & Stocklmayer, 2003; Bromme, Scharrer, Stadtler, Hömberg, & Torspecken, 2015; Hazen & Trefil, 2009). Here, scientific literacy is defined in terms of a deficit model, i.e., the notion that the general public lacks adequate knowledge of the sciences and therefore needs to be better educated on the basics of scientific processes and methods to understand how science works (Burns et al., 2003). Structured around a list of skills or attitudes, such as the American Association for the Advancement of Science's *Benchmarks for Science Literacy* (1993), this definition tends to be prescriptive and works toward bolstering science education standards.

A second broad category of definitions emphasizes the social structure of science (Burns et al., 2003; Durant, 1994), acknowledging that beyond the technical and cognitive concerns of science, there are ethical, political, social, and cultural understandings of science. In a recent study by the Pew Research Center (2015), citizens and scientists often disagree on a variety of science-related issues, including the safety of eating genetically modified foods, the need for vaccinations, the effects of human activity on climate change, and the theory of evolution. This category of definitions, then, is less about understanding the processes of science. Rather, it is more focused on the ability of an individual to read about, comprehend, and express an opinion on scientific matters (Miller, 1983).

Rennie and Stocklmayer (2003) write: "People selectively filter and restructure scientific information into a form they [find] personally meaningful and useful...the 'public' simply do not understand science on science's terms, but on their own terms. This includes understanding scientific ideas...but extends much further into issues of understanding risk, pride in local understandings, and cultural and societal values" (p. 765). In a 2009 editorial in *Science*, B. Alberts goes straight to the heart of the issue:

> Vast numbers of adults fail to take a scientific approach to solving problems or making judgments based on evidence. Instead, they readily accept simplistic answers to complicated problems that are confidently espoused by popular talk-show hosts or political leaders, counter to all evidence and logic. Most shocking to me is the finding that many college-educated adults in the United States see no difference between scientific and nonscientific explanations of natural phenomena such as evolution (p. 437).

Along these lines, the NRC's *National Science Education Standards* offered its definition of scientific literacy: "the knowledge and understanding of scientific concepts and processes required for personal decision making, participation in civic and cultural affairs, and economic productivity" (1996, p. 22). Similarly, the American Association for the Advancement of Science states that scientific literacy involves:

> Being familiar with the natural world and respecting its unity; being aware of some of the important ways in which mathematics, technology, and the sciences depend upon one another; understanding some of the key concepts and principles of science; having a capacity for scientific ways of thinking; knowing that science, mathematics, and technology are human enterprises, and knowing what that implies about their strengths and limitations; and being able to use scientific knowledge and ways of thinking for personal and social purposes.
> **American Association for the Advancement of Science. (1989), p. 20**

Tying these two categories of definitions together, the Organization for Economic Cooperation and Development's (OECD) *Programme for International Student Assessment (PISA) 2015 Assessment and Analytical Framework* offers a synthesized definition of scientific literacy to consider:

> Scientific literacy is the ability to engage with science-related issues, and with the ideas of science, as a reflective citizen. A scientifically literate person is willing to engage in reasoned discourse about science and technology, which requires the competencies to: **Explain phenomena scientifically**—recognise, offer and evaluate explanations for a range of natural and technological phenomena; **Evaluate and design scientific inquiry**—describe and appraise scientific investigations and propose ways of addressing questions scientifically; [and] **Interpret data and evidence scientifically**—analyse and evaluate data, claims and arguments in a variety of representations and draw appropriate scientific conclusions.
> **Organization for Economic Cooperation and Development [OECD]. (2016), p. 20**

Rychen and Salganik (OECD, 2003) wrote that scientific literacy is "an understanding of how [a knowledge of science] changes the way one can interact with the world and how it can be used to accomplish broader goals" (p. 10); they continue by acknowledging the role of information seeking in understanding the sciences.

SCIENTIFIC LITERACY AND INFORMATION LITERACY

Scientific literacy relates to a general ability to find and use information to understand science within a larger context; this general ability is commonly referred to as *information literacy*. Information literacy is defined by the Association of College and Research Libraries (ACRL) as "the set of integrated abilities encompassing the reflective discovery of information, the understanding of how information is produced and valued, and the use of information in creating new knowledge and participating ethically in communities of learning" (ACRL Board, 2016; Paragraph 4). The ACRL *Framework for Information Literacy for Higher Education* recognizes an interconnected core of foundational concepts, or frames, that serve as the conceptual underpinnings of information literacy. These are Authority is Constructed and Contextual; Information Creation as a Process; Information Has Value; Research as Inquiry; Scholarship as Conversation; and Searching as Strategic Exploration. The concepts extend beyond traditional competencies or skills required of an information literate individual, acknowledging not just the behavioral aspects of information literacy but also the cognitive, affective, and metacognitive aspects of learning.

Multiple frameworks exist for assessing scientific literacy, including the NRC's *Taking Science to School* (2007) proficiencies for science learning; the NRC's *Framework for K–12 Science Education* (2012); and the OECD's *PISA 2015 Science Framework* (2016). Porter et al. (2010) reflect that "Developing information literacy (IL) and [scientific literacy] SL requires similar skills and cognitive abilities; integrating these literacies is both efficient and meaningful" (p. 536). Thus, comparing and merging these scientific literacy and information literacy frameworks allows us to see the convergences of these two related domains.

One of the clearest areas of overlap relates to the process of scientific inquiry. ACRL's Research as Inquiry frame states: "Research is iterative and depends upon asking increasingly complex or new questions whose answers in turn develop additional questions or lines of inquiry in any field" (2016,

Paragraph 9). This aligns closely with NRC's practice dimensions of scientific literacy "asking questions and defining problems" (2012) and can be seen in the *PISA 2015 Science Framework*, which challenges students to "describe and appraise scientific investigations and propose ways of addressing questions scientifically" (p. 25). Inquiry, the act of seeking information, is a process that focuses on open or unresolved questions within a discipline or between disciplines. Students conducting scientific research need to know how to develop a hypothesis or research question that addresses a gap in knowledge; they also need to know how to develop an appropriate research design to address that question. There is another important aspect to this: Students need to understand research design not just for their own research, but also so that they are able to understand and evaluate the research of others.

Similarly, both scientific literacy and information literacy acknowledge the notion that scholarship is a conversation: "Communities of scholars, researchers, or professionals engage in sustained discourse with new insights and discoveries occurring over time as a result of varied perspectives and interpretations" (ACRL Board, 2016; Paragraph 10). This parallels the NRC's third scientific proficiency strand: Understanding the nature and development of scientific knowledge. This proficiency focuses on science as a way of knowing—a way of comprehending "how scientific knowledge is developed over systematic observations across multiple investigations, how it is justified and critiqued on the basis of evidence, and how it is validated by the larger scientific community" (2007, p. 251). Thus, scholarship is not an isolated activity but rather a process that needs to be understood in context over time; new knowledge is always entering the scholarly domain and challenging what we know. Students need to reject simple answers to complex problems and acknowledge and recognize the perspectives that each researcher brings to the scholarly conversation.

Two information literacy frames indirectly referenced in scientific literacy are Information Creation as a Process and Searching as Strategic Exploration. Information Creation as a Process is described by ACRL thusly: "Information in any format is produced to convey a message and is shared via a selected delivery method. The iterative processes of researching, creating, revising, and disseminating information vary, and the resulting product reflects these differences" (2016, Paragraph 7). This is tangential to NRC's fourth scientific proficiency strand of participating productively in scientific practices and discourse, which focuses on constructing and presenting scientific models and explanations via written arguments, graphs, and physical models (2007). ACRL's Searching as Strategic Exploration is defined: "Searching for information is often nonlinear and iterative, requiring the evaluation of a range of information sources and the mental flexibility to pursue alternate avenues as new understanding develops" (ACRL Board, 2016; Paragraph 11). This concept is embedded throughout the *PISA 2015 Science Framework*, particularly in the discovery of scientific arguments and evidence from multiple different sources.

Despite the overlaps, there are two foundational concepts in information literacy that are not reflected in the current frameworks for scientific literacy:

- Authority is Constructed and Contextual: "Information resources reflect their creators' expertise and credibility, and are evaluated based on the information need and the context in which the information will be used. Authority is constructed in that various communities may recognize different types of authority. It is contextual in that the information need may help to determine the level of authority required" (ACRL Board, 2016; Paragraph 6).
- Information Has Value: "Information possesses several dimensions of value, including as a commodity, as a means of education, as a means to influence, and as a means of negotiating and understanding the world. Legal and socioeconomic interests influence information production and dissemination" (Paragraph 8).

For the former, the notion of authority in evaluating scientific arguments and evidence is crucial to understanding the credibility of the argument, but it needs to be balanced with understanding that different communities of practice value different authorities and that students need to be aware of which voices may be marginalized through formal structures of authority. For the latter, information—as intellectual property—both impacts and is impacted by financial, social, and/or civic gains. For example, it is generally believed that the production of scientific knowledge is designed to filter out errors and bias but students need to understand that, as a human activity, bias can enter the research process; therefore, they need to critically evaluate scientific arguments and bias in regard to sponsored research. These two concepts are crucial to the scientific process and demonstrate that, while there are many overlaps between scientific literacy and information literacy, these models work best when approached in tandem as complementary ways of understanding.

IMPROVING SCIENTIFIC LITERACY USING INFORMATION LITERACY SKILLS

Interdisciplinary approaches and problem-based learning have permeated all disciplines, with a growing realization that multiple literacies are necessary to fully understand a given subject. The current frameworks for both scientific literacy and information literacy discussed in this chapter embrace the trends of interdisciplinary and transdisciplinary research in the sciences and have the capacity and elasticity to reflect the complexity of this collaborative research. Below are some examples of interdisciplinary coursework topics that undergraduate students may commonly encounter.

- A philosophy course that discusses ethics in agriculture as communicated through popular media, including the growing globalization of food, the role of technology in agricultural production, and the rights and responsibilities of consumers and producers (Barnhill, Budolfson, & Doggett, 2016).
- An environmental studies course that explores current environmental issues through nature writing, poetry, and essays (Kobzina, 2010).
- An agricultural economics class that develops a marketing plan, including market analysis, SWOT (strengths, weaknesses, opportunities, threats) analysis, industry trends, and competitor analysis, to sell meat produced and processed by the Animal Sciences department (Dugan, 2008).
- A biology class that requires picking a current controversial life sciences topic, reading newspaper articles on the topic, researching the topic more in-depth, and writing a "letter to the editor" that highlights the issues of their topic without offering their own opinion (Berman, 2013b).

Beyond integration into scientific courses, introducing concepts of scientific literacy into courses addressing current topics provides students with models for analyzing popular science information from a more informed perspective. This practice increases their exposure to scientific topics and moves them away from dualistic thinking about complex scientific issues.

The intention of the ACRL *Framework for Information Literacy* is to foster collaboration among librarians and disciplinary faculty, acknowledging that information literacy cannot and does not solely belong within the realm of the library. At the same time, scientific literacy extends beyond pure scientific expertise into the information landscape in which science is negotiated, contextualized, and understood. The collaboration between librarians and instructors when teaching the convergence of scientific literacy and information literacy works to expand the students' understanding of science.

This partnership creates a pedagogical shift beneficial to all parties—librarians, instructors, and students—and enhances the transfer of knowledge. By integrating concepts of scientific literacy into their instruction sessions, librarians will further help contextualize the complex information environment for students, providing them with lifelong critical thinking skills and successful literacy in both science and information that will progress over time.

REFERENCES

ABET Engineering Accreditation Commission. (2015). *Criteria for accrediting engineering programs*. Retrieved from http://www.abet.org/wp-content/uploads/2015/10/E001-16-17-EAC-Criteria-10-20-15.pdf.

Alberts, B. (2009). Redefining science education. *Science, 323*(5913), 437. http://dx.doi.org/10.1126/science.1170933.

American Association for the Advancement of Science. (1989). *Project 2061: Science for all Americans*. Washington, DC: American Association for the Advancement of Science, Inc.

American Association for the Advancement of Science. (1993). *Benchmarks for science literacy*. New York: Oxford University Press.

American Association for the Advancement of Science. (2010). *AAAS vision and change in undergraduate biology education*. Retrieved from http://visionandchange.org/.

American Chemical Society. (2015). *Undergraduate professional education in chemistry: ACS guidelines and evaluation of procedures for bachelor's degree programs*. Retrieved from https://www.acs.org/content/dam/acsorg/about/governance/committees/training/2015-acs-guidelines-for-bachelors-degree-programs.pdf.

Association of American Colleges and Universities. (2015). *Falling short? College learning and career success*. Retrieved from https://www.aacu.org/leap/public-opinion-research/2015-survey-falling-short.

Association of College and Research Libraries [ACRL] Board. (2016). *Framework for information literacy for higher education*. Retrieved from http://www.ala.org/acrl/standards/ilframework.

Barnhill, A., Budolfson, M., & Doggett, T. (2016). *Food, ethics, and society: An introductory text with readings*. New York: Oxford University Press.

Berman, E. (2013a). Transforming information literacy in the sciences through the lens of e-science. *Communications in Information Literacy, 7*(2), 161–170. Retrieved from http://www.comminfolit.org/index.php?journal=cil&page=article&op=view&path%5B%5D=v7i2p161&path%5B%5D=175.

Berman, E. (2013b). Scientific literacy. In P. Ragains (Ed.), *Information literacy: Instruction that works* (2nd ed.) (pp. 217–228). Chicago: American Library Association.

Bromme, R., Scharrer, L., Stadtler, M., Hömberg, J., & Torspecken, R. (2015). Is it believable when it's scientific? How scientific discourse style influences laypeople's resolution of conflicts. *Journal of Research in Science Teaching, 52*(1), 36–57. http://dx.doi.org/10.1002/tea.21172.

Burns, T. W., O'Connor, D. J., & Stocklmayer, S. M. (2003). Science communication: A contemporary definition. *Public Understanding of Science, 12*(2), 183–202.

Dugan, M. (2008). Embedded librarians in an Ag econ class: Transcending the traditional. *Journal of Agricultural & Food Information, 9*(4), 301–309. http://dx.doi.org/10.1080/10496500802480342.

Durant, J. (1994). What is scientific literacy? *European Review, 2*(1), 83–89.

Harvard Transdisciplinary Research in Energetics and Cancer Center. (2017). *Definitions*. Retrieved from https://www.hsph.harvard.edu/trec/about-us/definitions/.

Hazen, R. M., & Trefil, J. (2009). *Science matters: Achieving scientific literacy*. New York: Anchor Books.
Institute of Physics. (2011). *The physics degree: Graduate skills base and the core of physics*. Retrieved from http://www.iop.org/education/higher_education/accreditation/file_43311.pdf.
Kobzina, N. (2010). A faculty-librarian partnership: A unique opportunity for course integration. *Journal of Library Administration, 50*(4), 293–314. http://dx.doi.org/10.1177/09636625030122004.
Lupton, M. (2008). Evidence, argument and social responsibility: First-year students' experiences of information literacy when researching an essay. *Higher Education Research & Development, 27*(4), 399–414. http://dx.doi.org/10.1080/07294360802406858.
Miller, J. D. (1983). Scientific literacy: A conceptual and empirical review. *Daedalus, 112*(2), 29–48. Retrieved from http://www.jstor.org/stable/20024852.
National e-Science Centre. (n.d.). *Defining e-science*. Retrieved from http://www.nesc.ac.uk/nesc/define.html.
National Research Council [NRC]. (1996). *National science education standards: Observe, interact, change, learn*. Washington, DC: The National Academy Press.
National Research Council [NRC]. (2007). Taking science to school: Learning and teaching science in grades K–8. Committee on science learning, kindergarten through eighth grade. In R. A. Duschl, H. A. Schweingruber, & A. W. Shouse (Eds.), *Board on science education, center for education. Division of behavioral and social sciences and education*. Washington, DC: The National Academies Press.
National Research Council [NRC]. (2009). *Transforming agricultural education for a changing world*. Washington, DC: The National Academies Press.
National Research Council [NRC]. (2010). *Research at the intersection of the physical and life sciences*. Washington, DC: The National Academies Press.
National Research Council [NRC]. (2012). *A framework for K-12 science education: Practices, crosscutting concepts, and core ideas*. Washington, DC: The National Academies Press. Retrieved from https://www.nap.edu/catalog/13165/a-framework-for-k-12-science-education-practices-crosscutting-concepts.
National Science Foundation. (2016). *Research areas*. Retrieved from https://www.nsf.gov/about/research_areas.jsp.
Organization for Economic Cooperation and Development [OECD]. (2003). In D. S. Rychen & L. H. Salganik (Eds.), *Definition and selection of key competencies, executive summary*. Göttingen, Germany: Hogrefe Publishing. Retrieved from http://www.oecd.org/pisa/35070367.pdf.
Organization for Economic Cooperation and Development [OECD]. (2016). *Programme for International Student Assessment (PISA) 2015 assessment and analytical framework: Science, reading, mathematic and financial literacy*. Paris: OECD Publishing. http://dx.doi.org/10.1787/9789264255425-en.
Pew Research Center. (2015). *Public and scientists' view on science and society*. Retrieved from http://www.pewinternet.org/2015/01/29/public-and-scientists-views-on-science-and-society/.
Porter, J. A., Wolbach, K. C., Purzycki, C. B., Bowman, L. A., Agbada, E., & Mostrom, A. M. (2010). Integration of information and scientific literacy: Promoting literacy in undergraduates. *CBE-Life Sciences Education, 9*(4), 536–542. http://dx.doi.org/10.1187/cbe.10.
Quinn, C., Burbach, M. E., Matkin, G. S., & Flores, K. (2009). Critical thinking for natural resource, agricultural, and environmental ethics education. *Journal of Natural Resources & Life Sciences Education, 38*, 221–227. http://dx.doi.org/10.4195/jnrlse.2009.0028.
Rennie, L. J., & Stocklmayer, S. M. (2003). The communication of science and technology: Past, present and future agendas. *International Journal of Science Education, 25*(6), 759–773. http://dx.doi.org/10.1080/09500690305020.
Vincent, S., & Focht, W. (2009). *Perspectives on environment program curricula and core competencies: A report of the Curriculum Committee of the Council of Environmental Deans and Directors*. Washington, DC: National Council for Science and the Environment.

CHAPTER 3

Designing Information Literacy Instruction for the Life Sciences

Katherine O'Clair
California Polytechnic State University, San Luis Obispo, CA, United States

INTRODUCTION

This chapter addresses designing instruction for information literacy in the life sciences and is ideal for those seeking to learn more about planning and developing information literacy instruction for students in these disciplines. It is intended to be practical in nature, providing insight into the important things to consider when designing, delivering, and assessing student learning experiences. The focus is on undergraduate education as it applies to the widest audience, yet much of what is discussed can be applied to graduate students and graduate-level programs. This chapter will describe the life sciences curriculum and characteristics of students in such programs and provide practice-based examples and approaches. Programmatic approaches for information literacy in the life sciences will be shared, and several tools and strategies used to design, deliver, and assess information literacy will be introduced. While there will always be subtle differences between the disciplines encompassed in the life sciences, this chapter will treat the life sciences in a general sense. The discipline-specific chapters that follow will provide further discussion of information literacy strategies for individual subjects within the life sciences.

THE LIFE SCIENCES CURRICULUM

To be effective instructors, librarians should have an awareness of the disciplinary curriculum for which they are providing information literacy instruction. Again, the information presented here is general in nature, and librarians will want to familiarize themselves with the specific curricula at their institutions. Such knowledge helps to align the information literacy with the curriculum, thus enhancing the experience for students and contributing to their success in learning.

The science curriculum is inherently different from other curricula such as business or the humanities. Much of science deals with abstract concepts and theories (e.g., natural selection, conservation of energy) that cannot be directly observed (Leonard, 1997). The life sciences curriculum is broad and encompasses a wide range of disciplines beyond those covered in this book. Courses in the life sciences majors typically contain a lecture component where the concepts and theories are taught. Some courses may have a complementary laboratory section in which students gain hands-on experience with the concepts they are learning in the lecture component. In addition, selected courses, especially those with which students commonly struggle, may contain a recitation section to provide additional time to address challenging or advanced concepts.

The requirements of the degree programs can vary from discipline to discipline and institution to institution, yet it is common for each to have specified program learning objectives that include knowledge of the discipline, as well as critical thinking and lifelong learning. Librarians should consult these learning objectives as they engage with faculty in the disciplines to design and plan information literacy activities. Additionally, some programs (e.g., human nutrition) may be accredited by outside organizations or governing bodies, and they may have competency guidelines for program graduates that include information literacy and related skills.

Lecture Courses

Many courses in the life sciences employ a lecture-based format of instruction to teach the theory. These lecture courses often utilize multiple lecture exams and a final exam to assess student learning, although some will integrate additional quizzes, papers, and projects. Recently, there has been a trend toward incorporating more active learning modalities, including problem-based learning, case studies, and computer-based activities, into the science curriculum (Handelsman et al., 2004).

Laboratory Sections

Courses in the life sciences typically have laboratory sections associated with them to provide students with hands-on experiences to learn the concepts covered in the lecture portion of the course. It is the ideal setting for students to develop the habits of mind and practices of a scientist, especially the ability to identify the characteristics of science and think scientifically (Leonard, 1997). These laboratory sections are held once or twice weekly for several hours to allow the time to complete the laboratory-based explorations of the concepts. They can be more accessible in terms of the availability of time to integrate additional activities. Moreover, these courses often require students to complete laboratory reports using the IMRAD format (Introduction, Methods, Results, and Discussion). Students are typically required to consult and cite authoritative sources in the Introduction and Discussion sections, and this provides a meaningful opportunity for integrating information literacy concepts and skills and to produce work that can be directly assessed to measure their learning.

Lower-Division Courses

In the curriculum for life sciences disciplines, many lower-division courses serve as prerequisites for upper-division courses in those majors. Consequently, these lower-division courses offer an opportunity to introduce foundational skills that will be beneficial and necessary as students complete the subsequent, more advanced courses in their majors. They can also serve as support courses for majors in other disciplines, such as biomedical engineering or psychology. In addition, many institutions require all students regardless of major to take a science course as part of their general education requirements. Life sciences courses, especially introductory biology courses, are quite popular as an option to fulfill the science general education requirement. In fact, some institutions offer a nonmajor biology course designed to meet the needs and demands of this population. This is an important consideration to remember when planning information literacy instruction for these courses, as not all students will be equally facile and engaged with the concepts they are learning.

Upper-Division Courses

Courses at the upper-division level build on the content introduced in lower-division courses, incorporating more advanced concepts and topics. Upper-division courses are very specific to the major within the discipline and often contain more content than there is time to fully integrate, especially as scientific knowledge continues to grow at a rapid pace (Leonard, 1997). Consequently, these courses provide fewer opportunities to incorporate additional learning objectives such as information literacy instruction. Despite this constraint, they provide excellent opportunities for students to practice the skills they have acquired in lower-division courses, helping them move toward proficiency. In many cases, students can be asked to incorporate what they have learned previously, with moderate extensions of learning, without the need to devote large blocks of class time for additional instruction. Such courses also lend themselves well to an embedded model of librarian involvement.

Required Versus Elective

The curriculum in each major in the life sciences contains a number of required courses that students must complete to earn their degrees. These will vary among the majors, but there is likely to be some overlap. The elective courses allow students to shape their majors to meet their specific needs and interests. When planning information literacy instruction in these majors, it is best to target the required courses, as this will reach the greatest number of students in the degree program. The elective courses are best suited for providing additional opportunities for students to practice the information literacy skills they have learned in the required courses, which will increase the likelihood of achieving proficiency toward the end of the degree program.

CHARACTERISTICS OF STUDENTS IN THE LIFE SCIENCES

Most students taking life sciences courses will be majors in one of the life sciences disciplines previously mentioned. Similar to other disciplines, students in these programs have a variety of ambitions for the future including further graduate-level study, professional programs, and immediate employment in their chosen fields and industries. Students pursuing further study may be seeking advanced degrees to pursue careers as bench or field scientists. Others may be seeking opportunities to teach, either in the K–12 educational system or in higher education. Those pursuing professional programs typically go into medical, dental, and veterinary doctoral programs that are highly selective and competitive. A large number of students go directly into the workforce, especially those in the agricultural and environmental sciences programs.

Students in life sciences courses, particularly those at the lower-division level, may also be nonmajors completing requirements for general education or supported degree programs (e.g., biomedical engineering, psychology). Typically, these students have had less exposure to the discipline, and they may not be as invested as students in the major. This aspect of students in life sciences courses is important to mention, as the design and the delivery of the instruction may need to be adapted to meet the situational needs of all students in the course.

COLLABORATING WITH INSTRUCTORS

To be successful in integrating information literacy, librarians must collaborate with instructors. Instructors are often eager to find ways to familiarize students with the scholarly literature and to get them to use more appropriate and relevant sources in their assignments. Despite this eagerness, the demands on instructional faculty, particularly at research-intensive institutions, sometimes make it difficult for them to give their full attention to information literacy. As mentioned before, there is often more content than time for a course, leaving limited options for additional content or instruction. This should not be a source of discouragement, rather something to consider in negotiating with instructors to incorporate information literacy instruction and activities into their courses.

At research-intensive institutions, graduate teaching assistants can be willing and reliable partners in integrating information literacy instruction and activities. They often teach the laboratory sections of courses, where more time may be available to add content or instruction. In addition, these graduate assistants often are interested in learning more about information literacy themselves, as they need advanced skills to achieve their own educational aims. Moreover, these individuals are likely to be the instructional faculty of the future, so involving them in collaborations to advance information literacy skills and abilities in students now will establish a practice of working with librarians that they will carry with them into their future careers.

KEY OPPORTUNITIES FOR INTEGRATING INFORMATION LITERACY

The life sciences curriculum offers ample opportunities to integrate information literacy into existing course activities. Some examples of this include laboratory reports, research papers, capstone projects, and undergraduate research experiences.

Laboratory Reports

Laboratory courses are ubiquitous in the life sciences, as many of the topics require hands-on experience with and investigation of the concepts. The laboratory report is a popular artifact of student learning in these courses, and most instructors require students to consult outside information sources as part of this process. As such, this provides a relevant and meaningful opportunity for information literacy. In fact, the laboratory report can be used to integrate several of the frames from the Association of College and Research Libraries' 2016 *Framework for Information Literacy for Higher Education* (ACRL Board, 2016), which will be discussed later in this chapter.

Research Papers

Another popular artifact of student learning in life sciences is the research or term paper. These will often be included as part of lecture courses in addition to the exams. At some institutions, select life sciences courses may be writing-intensive, allowing students to fulfill both the curricular requirement and the institution's writing requirement. For the most part, instructors will expect students to use the scholarly, peer-review literature as sources in these papers, and this provides an excellent opportunity to introduce and reinforce how to discern different types of literature sources and the concept of peer review.

Capstone Projects

The capstone course or "senior" seminar is common in the life sciences curriculum, and it provides another great opportunity for engagement. These courses are intended to "bring everything together" and allow students to demonstrate their proficiency with synthesizing concepts essential to the discipline of the major. Written reports, oral presentations, posters, and portfolios are all common products of student learning in capstone courses. Consequently, this is an excellent opportunity for students to demonstrate proficiency in information literacy and its related skills, knowledge, and abilities. Despite this, in reality, it is common for instructors to reach out to librarians to ameliorate poor information research skills in students who are expected to perform at a much higher level of competency but have not been given adequate opportunities for learning and practice.

Undergraduate Research Experiences

Undergraduate research experiences have become more popular in recent decades, and these provide an excellent opportunity to work directly with students to support their research endeavors. According to F. Farmer, Director of Research Development, Arizona State University, funding agencies such as the National Science Foundation and the National Institutes of Health are showing an increasing interest in including undergraduate students in sponsored research projects (F. Farmer, Personal communication, December 27, 2016). These experiences often include a culminating project, which is often presented orally or in written form. Some institutions have self-published undergraduate research journals in which students submit their scholarship for peer review and publication. Librarians are important partners here, as they can provide group instruction and individualized assistance to student participants with their projects and assist them with the dissemination of their results through posters, presentations, and papers.

INFORMATION LITERACY COMPETENCIES

The information literacy skill set desired in students is another important aspect to consider when designing information literacy for life sciences. Both the *Information Literacy Standards for Science and Engineering/Technology* (ALA/ACRL/STS Task Force on Information Literacy for Science and Technology, 2006) and the *Framework for Information Literacy for Higher Education* (ACRL Board, 2016) can be used to guide the development of information literacy competencies for students in the life sciences. Neither of these documents is designed to be prescriptive, rather they should be used to identify and develop the criteria for information literacy competence in life sciences students. Instructional faculty should be included in these discussions, as they can provide insight and specific details about the information literacy skills they expect of students on completion of their respective degree programs.

Information Literacy Standards for Science and Engineering/Technology

The *Information Literacy Standards for Science and Engineering/Technology* derive from the now-superseded *Information Literacy Competency Standards for Higher Education* (ACRL, 2000) and specifically address the specialized characteristics of the science, engineering, and technology disciplines. Each of the standards includes performance indicators and associated objectives to describe the skills and abilities desired of information literate students in science-related disciplines. Although the professional community has moved toward a threshold concepts-based approach for information literacy competency, it is helpful to use elements of these still-adopted standards to guide the planning and design of information literacy in the life sciences, especially as students engage more with the data and visualized information that prevails in the life sciences disciplines. It will continue to be important for students in the life sciences to be able to

- specify the type and amount of information needed;
- gather the appropriate information in a productive manner;

- critically evaluate information and sources and modify the information seeking strategy as needed;
- use information in an ethical and legal manner; and
- engage in lifelong learning to stay up-to-date with emerging trends and developments in the field (ALA/ACRL/STS Task Force on Information Literacy for Science and Technology, 2006).

Critical evaluation of information and its sources is also important, especially considering the recent onslaught of "fake news" that has quickly dominated the social media landscape. The last outcome listed is likely the most important, as professionals in the life sciences (e.g., physicians, scientists) are expected to stay continuously engaged with information relevant to their professions throughout their careers.

Framework for Information Literacy for Higher Education

The ACRL's (2016) *Framework for Information Literacy for Higher Education* provides a new and flexible way in which to employ information literacy in higher education. The Framework emphasizes a "dynamic and often uncertain information ecosystem," the role that students play in "creating new knowledge," and the importance of "metacognition, or critical self-reflection." It focuses on *threshold concepts*, which represent the idea that once students have grasped the knowledge fully they are transformed; the knowledge cannot be lost or undone; and they will not retreat to their previous, limited state of understanding (Meyer & Land, 2006). The ultimate goal is a shift forward in their understanding of a complex idea.

One of the most important aspects of the *Framework for Information Literacy for Higher Education*, and one that is fundamentally different from the previously sanctioned *Information Literacy Competency Standards for Higher Education*, lies in how it should be employed. The Framework is not intended to be a directive, rather it allows stakeholders to determine how to best use it based on their own needs at their respective institutions (ACRL Board, 2016).

The *Framework for Information Literacy for Higher Education* can be viewed through the lens of the life sciences, and the following provides a brief summary of some of the important aspects of each frame.

Authority Is Constructed and Contextual

Scientists and researchers who publish in scholarly, peer-reviewed journals are the long-standing, recognized authority in the life sciences. While there are always exceptions, students should recognize this standard of authority within the discipline and maintain a healthy skepticism about using sources for which authority cannot be verified. Information in the life sciences often requires a high level of authority. For instance, medical information should come from individuals who have been properly trained, credentialed, and vetted.

Information Creation as a Process

In the sciences, information creation is grounded in the scientific method, which includes "(a) systematic observation, measurement, and experimentation; (b) induction and the formulation of hypotheses; (c) the making of deductions from the hypotheses; (d) the experimental testing of the deductions; and (if necessary) (e) the modification of the hypotheses" (Scientific Method, 2016). The predominant form for delivery of information in the life sciences is the scholarly, peer-reviewed article. Despite both of these longstanding processes, new models of information creation and publication are emerging as a result of advances in technology (e.g., open peer review).

Information Has Value

The high cost of journals in the life sciences disciplines is one of the most prevalent examples of this concept. Related to this is the notion of free access to taxpayer-funded research, which has been available at an increasing rate for nearly a decade. In addition, students should understand the importance of giving credit where due and the value of their own contributions to the information landscape. This latter aspect is critical for underrepresented minorities in the life sciences, who may experience challenges to professional advancement disproportionate to other groups of students (F. Farmer, Personal communication, December 27, 2016).

Research as Inquiry

The purpose of scientific research is to investigate unanswered questions and gaps in the existing knowledge within the field of study. One of the fundamental roles of scientific study is to generate further questions for inquiry and investigation. In the sciences, the scholarly literature is the forum for the debate and discussion of knowledge within the scientific community. Students should recognize the importance of using appropriate sources to draw conclusions and synthesize ideas in the life sciences.

Scholarship as Conversation

Scientists have traditionally communicated results of research and debated issues through the scholarly, peer-reviewed literature. This long-standing practice is evolving as new and social media gain traction as a means of communicating scientific information. Students play an important part in this construct as emerging professionals, and they should begin to view themselves and engage as participants in this conversation in preparation for their future professional work.

Searching as Strategic Exploration

Students must know the important sources of information in the life sciences and then search them effectively to locate the needed information, matching their information needs with the best approaches and tools. As such, it is important to utilize the most effective tool, to use an appropriate search strategy, and to select a relevant information source. As with other disciplines this process is iterative and requires a high level of persistence and curiosity. Students in the life sciences are likely to be more facile in this way, as these disciplines often attract those drawn to inquiry.

DESIGNING WITH THE OBJECTIVE–ACTIVITY–ASSESSMENT APPROACH

This next section addresses a systematic, assessment-based approach for information literacy in the life sciences. The topic of assessment-based approaches for student learning is broad and deep and includes much more than can be covered in this chapter alone. What follows will "scratch the surface" of this topic, highlighting the important aspects and providing practical guidance.

The Objective–Activity–Assessment Approach (Fig. 3.1) includes the following steps: develop learning objectives, design learning activities, and assess student learning (note: delivery of instruction is part of the sequence but will not be addressed in this chapter). Assessment is defined as "the ongoing process of
- establishing clear, measurable expected outcomes of student learning;
- ensuring that students have sufficient opportunities to achieve those outcomes;

Figure 3.1 Steps in the Objective–Activity–Assessment Approach.

- systematically gathering, analyzing, and interpreting evidence to determine how well student learning matches our expectations; and
- using the resulting information to understand and improve student learning" (Suskie, 2009, p. 4).

The Teaching Tripod Approach employs a similar, straightforward method for designing information literacy instruction that matches well with the definition above. It incorporates expected learning outcomes, activities, and assessment into the instructional design process by examining what we expect students to learn, how they will learn it, and how we know they have learned, respectively (Kaplowitz, 2014).

Developing Learning Objectives

Learning objectives (also known as learning outcomes) are essential for effective learning. They help to articulate what students should be able to do as a result of the instruction and consequently aid in designing more effective instruction planning, activities, and assessments (Gronlund, 2000). When developing learning objectives, carefully consider what students should learn and be able to accomplish from the instruction. The revised Bloom's Taxonomy (Anderson et al., 2001) is very helpful for writing action-based learning objectives and identifying the appropriate cognitive level (Fig. 3.2).

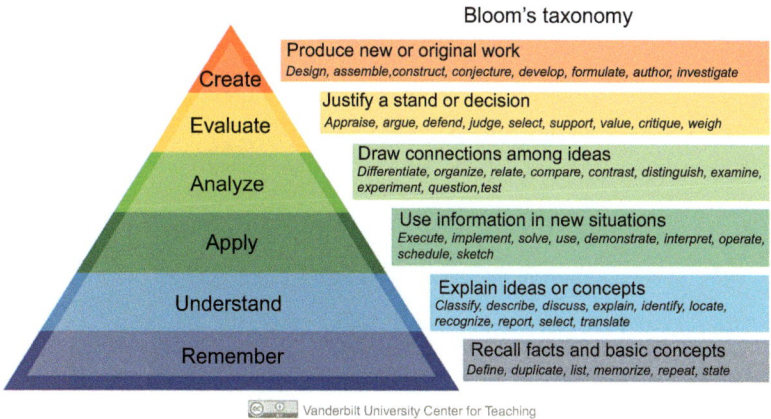

Figure 3.2 Revised Bloom's Taxonomy showing cognitive levels and associated verbs CC by 2.0, Vanderbilt University Center for Teaching. (2001). *Bloom's taxonomy (licensed under Creative Commons 2.0 [no changes made])*. Retrieved from https://www.flickr.com/photos/vandycft/29428436431.

A basic formula for creating a learning objective is:

Students will be able to + Action (verb) + Skill/Knowledge/Ability

For instance, in an environmental sciences course, students should be familiar with examples of gray literature relevant to that subject area (e.g., government documents such as US Forest Service General Technical Reports).

Learning Objective = Students will be able to name two relevant examples of gray literature relevant to the environmental sciences

This learning objective would fall in Level 1 (Remember).

Moving to the next level (Understand), students should understand how gray literature sources are important to the field of environmental sciences.

Learning Objective = Students will be able to explain the role of gray literature sources in the environmental sciences

Writing learning objectives can be challenging, especially at first. With practice, however, it becomes easier, and there are a number of books, articles, and web-based resources available to provide guidance.

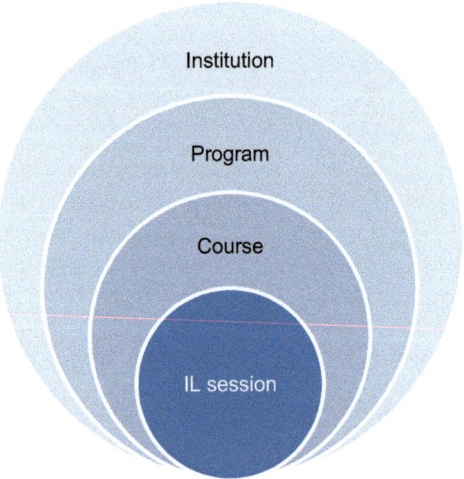

Figure 3.3 The relationship of an information literacy (IL) session to learning outcomes at the various levels. *Adapted with permission Fabbi, J. (October 2014). Creating context for information literacy: Best practices for learning and assessment. In* WASC senior college and university commission, retreat on core competencies: Critical thinking and information literacy, October 16–17, 2014, Oakland, CA *(Unpublished)*.

Information literacy learning objectives should be aligned with those at the course, program, and institution levels (Fig. 3.3). The learning objectives created for the information literacy component of the course should be connected to what students are expected to learn as part of the course itself. Collectively, these objectives should contribute to the achievement of the program-level and, ultimately, the institution-level objectives. Such alignment ensures that the information literacy instruction is value added and intentional (O'Clair, in press).

Designing Learning Activities

Once the learning objectives have been established, the learning activities can be designed and planned to ensure that the concepts are reaching students and that the students are given adequate opportunities to learn them. The learning activities should always assist students in achieving the expected learning objectives. If the learning objectives change over time, the learning activities will need to be revised to reflect the new updated objectives, and vice versa.

Designing active learning experiences gives students the greatest chances for successful learning, because they will be engaged in the learning process and will develop a deeper understanding of the material that will be retained longer (Leonard, 1997). Active and problem-based learning approaches are particularly well suited for the life sciences, as students are familiar with employing concepts in the laboratory, and the curriculum provides a wealth of real-world examples with which to work. For example, many instructors in the life sciences want their students to be able to locate and read the scholarly literature in their respective fields. A problem-based learning experience in which a research team (in this case, a group of students) must find and retrieve several articles on habitat utilization by migratory songbirds during stopover (in this case, the research topic) to identify existing knowledge and gaps (in this case, synthesize the information) would work well here.

One of the most important aspects of designing learning activities involves ensuring that students can articulate the purpose of the activity, the tasks associated with the activity, and the measures used to determine successful completion. Transparent assignment design helps to create learning opportunities that are student centered, making it easier for students to grasp the concepts they are expected to learn and why they are important. The University of Nevada, Las Vegas, has developed a freely available checklist for designing transparent assignments for use in developing learning activities and assignments for information literacy instruction (Winkelmes, 2016).

Assessing Student Learning

Assessment has gained considerable popularity in recent decades, as institutions strive to measure their value and demonstrate student learning. Libraries should engage in meaningful assessment to demonstrate their value to the institution and to measure their contributions to student learning, especially in the area of information literacy (Oakleaf, 2011). The process of assessing student learning can be complex and challenging, yet there are a number of straightforward approaches that can be incorporated using formative and summative assessment methods.

Formative assessment measures student learning along the way, providing opportunities to make adjustments to the teaching and learning to ensure students are progressing as expected. This is beneficial to the students, who can use the formative assessment activity to self-regulate their understanding of the concepts, as well as to the instructor, who can identify whether students are grasping the concepts being taught. Classroom Assessment Techniques (CATs) are widely used as a type of formative assessment (Angelo & Cross, 1993) and can be used in a variety of ways and at multiple points during instruction. For instance, using the prior example about gray literature sources, an online poll could be incorporated mid-way through the instruction session to ensure that students are able to identify gray literature sources of information, and then adjust accordingly if they fall short of the expected results. Formative assessment is beneficial to students because it provides early identification of and intervention for students who are struggling with the concepts, as well as multiple chances to learn and move toward success in achieving the learning objectives. In addition, it helps students be more successful in adequately completing the culminating course activities such as the summative assessment examples of term papers and final exams.

Summative assessment measures student learning on completion of the learning activities or the course, and it is usually focused on the intended learning outcomes (Suskie, 2009). In life sciences courses, this generally comes in the form of a culminating assignment, such as a term paper or presentation, or a portfolio as part of a capstone course. Rubrics are often used to assess student learning through these products, as they provide detailed characteristics of performance that are scaled and can be applied subjectively (Booth, 2011). In information literacy instruction, many of the same methodologies used in formative assessment can be applied (Booth, 2011). Self-reflection, as one example of summative assessment, can be effective, especially if it is employed holistically in an artifact such as a Researcher's Memo, in which students discuss and evaluate their own learning throughout the entire experience. While summative assessment occurs at the end of the learning experience, it should not be confused with evaluation. The emphasis should be on how well the student has learned the content, not how they rated the experience or its delivery.

Assessment measures generally fall into two categories: direct and indirect. "Direct evidence of student learning is tangible, visible, self-explanatory, and compelling evidence of exactly what students have or have not learned" (Suskie, 2009, p. 20). Examples of direct evidence include research papers and projects, capstone experiences, and students' self-reflection on their learning. "Indirect evidence consists of proxy signs that students are probably learning" (Suskie, 2009, p. 20). Course grades, graduation rates, and student perceptions are common types of indirect evidence of student learning. Direct measures should be used as often as possible because they provide actual evidence of learning from the students' work and experiences (e.g., from self-reflection).

Programmatic Approaches for Information Literacy in the Life Sciences

One of the first steps toward developing a more programmatic approach to information literacy instruction involves taking an inventory of established and potential opportunities for instruction. Librarians need to know the aforementioned characteristics of the life sciences curricula to determine the following questions:
- In which courses do students currently receive information literacy instruction?
- Where do those courses fall in the curriculum?
- What courses are students required to complete for the major?
- When do students take those required courses?
- Are there opportunities to integrate additional instruction?
- Are there concentrated areas of information literacy instruction (or gaps) in the curriculum?

The next step involves determining the sequence of courses in which students will receive information literacy instruction. Librarians should work closely with academic departments in this activity to gain an understanding of how students typically move through the coursework in the major. In this process, determine the best possible courses in which to include information literacy instruction or activities. Instructors and departments need to agree to allocate time in their courses and programs to integrate information literacy, and this often must be negotiated with flexible and creative solutions.

The final step ensures that students will have sufficient opportunities to achieve the intended learning objectives. This involves identifying the courses in which the learning objectives will be addressed; determining where and when the corresponding concepts and skills will be introduced and practiced; and selecting the courses in which students are expected to demonstrate proficiency and achievement of the desired learning objectives.

RECOMMENDATIONS AND PRACTICAL ADVICE

Many librarians charged with information literacy instruction in the life sciences serve academic units with large numbers of students, and this is especially the case at research-intensive institutions. Regardless of the institution's size, creating instruction that is scalable will help to optimize the instructional efforts to achieve the widest reach. While not all information literacy efforts can be scaled easily, it is important to consider how to design and deliver the instruction to reach the greatest number of students without losing its effectiveness. Piloting a new instructional activity on a smaller scale to start can provide the insight needed to adjust it to reach a wider number of students. In many cases, the materials and approaches will already be developed, and these can be adapted to fit a larger group or applied in an additional discipline.

Planning and designing instruction that can be adapted to withstand changes in personnel, curriculum, and priorities will help to ensure sustainability of your instructional endeavors. Many instructors teach a variety of courses from term-to-term, and having multiple sections of a course is a common occurrence, especially for foundational courses in the life sciences. Thus, collaboration with several different instructors may be needed to integrate the instruction that has been designed. One of the benefits of using the Objective–Activity–Assessment Approach outlined above is that it helps to articulate and demonstrate the value of information literacy in the curriculum, which will be essential over time as priorities of both the department and the library evolve and change.

It is important to be intentional and strategic when planning information literacy instruction in the life sciences. The Information Literacy Instruction Planning Worksheet can assist librarians with planning information literacy instruction in a more intentional and strategic manner (O'Clair, in press). The process of reflecting on both the successes and failures with information literacy instruction will also help. The process of planning in an intentional and strategic way can be difficult, so it is important not to let "perfect" get in the way of "good." Sometimes "good enough" really is good enough and the most practical.

Planning, designing, and delivering information literacy instruction is a time-intensive endeavor. Frequently, the strategies and successes of our colleagues in their instructional efforts are shared with little mention of the painstaking and lengthy amount of work that went into the process. While it is beneficial to focus on success, it is important to understand that these endeavors often do not pay off immediately and certainly not without significant time and effort. Set realistic goals and expectations for all stakeholders. Celebrate the successes and learn from what did not go as expected. Sometimes we gain the most insight from those things that did not turn out the way we intended. Give this process the time it needs to develop and mature in the way that will be the best for all involved.

Successful integration of information literacy into the life sciences curriculum relies heavily on collaboration and partnership with the academic department and its instructors. In fact, because of the compressed curriculum in the life sciences disciplines and the subjective nature of the learning activities, buy-in from the department and the instructors is required to best match the instructional intervention with the need and to negotiate for valuable class time to deliver the instruction. Librarians should be flexible and creative in identifying the best strategy for integrating information literacy activities and instruction into both the specific course and the curriculum. One-size-fits-all rarely works in reality, and librarians should be encouraged and empowered to determine what is best for their situations and then design and deliver accordingly.

CONCLUSION

Designing information literacy for the life sciences is an exciting and engaging endeavor, and this chapter has examined this process by providing an overview of the important aspects to consider. Whether it is a one-shot session or a fully integrated information literacy program, the design, development, and delivery of the instruction requires librarians to be familiar with the concepts presented here. With this arsenal of information, librarians can begin to establish the situation-specific teaching and learning practices that will provide a satisfying and meaningful educational experience for all involved.

RECOMMENDED RESOURCES

Booth, C. (2011). *Reflective teaching, effective learning: Instructional literacy for library educators.* Chicago: American Library Association.

Bowles-Terry, M., & Kvenild, C. (2015). *Classroom assessment techniques for librarians.* Chicago: Association of College and Research Libraries.

Bravender, P., McClure, H., & Schaub, G. (Eds.). (2015). *Teaching information literacy threshold concepts: Lesson plans for librarians.* Chicago: Association of College and Research Libraries.

Broussard, M. S., Hickoff-Cresko, R., & Oberlin, J. U. (2012). *Snapshots of reality: A practical guide to formative assessment in library instruction.* Chicago: Association of College and Research Libraries.

Hopkins, E. (2012). Life and health sciences. In K. O'Clair & J. Davidson (Eds.), *The busy librarian's guide to information literacy in science and engineering* (pp. 35–46). Chicago: Association of College and Research Libraries.

Kaplowitz, J. (2014). *Designing information literacy instruction: The teaching tripod approach.* Lanham, MD: Rowman & Littlefield.

Suskie, L. (2009). *Assessing student learning* (2nd ed.). San Francisco: John Wiley & Sons.

ACKNOWLEDGMENTS

This chapter is dedicated in loving memory of Dr. Cari Lynn Finney Ouderkirk, who believed in me and taught me to believe in myself. Special thanks to Faye Farmer and Catherine Hillman for their thoughtful review of and detailed guidance on this chapter.

REFERENCES

ALA/ACRL/STS Task Force on Information Literacy for Science and Technology. (2006). *Information literacy standards for science and engineering/technology.* Retrieved from http://www.ala.org/acrl/standards/infolitscitech.

Anderson, L. W., Krathwohl, D. R., Airasian, P. W., Cruikshank, K. A., Mayer, R. E., Pintrich, P. R., et al. (Eds.). (2001). *A taxonomy for learning, teaching, and assessing: A revision of Bloom's taxonomy of educational objectives* (Abridged ed.). New York: Addison Wesley Longman, Inc.

Angelo, T. A., & Cross, K. P. (1993). *Classroom assessment techniques: A handbook for college teachers.* San Francisco: Jossey-Bass.

Association of College and Research Libraries [ACRL]. (2000). *Information literacy competency standards for higher education*. Retrieved from http://www.ala.org/acrl/sites/ala.org.acrl/files/content/standards/standards.pdf.

Association of College and Research Libraries [ACRL] Board. (2016). *Framework for information literacy for higher education*. Retrieved from http://www.ala.org/acrl/standards/ilframework.

Booth, C. (2011). *Reflective teaching, effective learning: Instructional literacy for library educators*. Chicago: American Library Association.

Fabbi, J. (October 2014). Creating context for information literacy: Best practices for learning and assessment. Unpublished. In *WASC senior college and university commission, retreat on core competencies: Critical thinking and information literacy, October 16–17, 2014, Oakland, CA*.

Gronlund, N. E. (2000). *How to write and use instructional objectives* (6th ed.). Upper Saddle River, NJ: Prentice-Hall, Inc.

Handelsman, J., Ebert-May, D., Beichner, R., Bruns, P., Chang, A., DeHaan, R., et al. (2004). Scientific teaching. *Science, 304*(5670), 521–522.

Kaplowitz, J. (2014). *Designing information literacy instruction: The teaching tripod approach*. Lanham, MD: Rowman & Littlefield.

Leonard, W. H. (1997). How do college students learn science? In E. Siebert, M. Caprio, & C. Lyda (Eds.), *Methods of effective teaching and course management for university and college science teachers* (pp. 5–20). Dubuque: Kendall/Hunt.

Meyer, J. H. F., & Land, R. (2006). Threshold concepts and troublesome knowledge: An introduction. In J. H. F. Meyer, & R. Land (Eds.), *Overcoming barriers to student understanding: Threshold concepts and troublesome knowledge* (pp. 3–18). New York: Routledge.

Oakleaf, M. (2011). Are they learning? Are we? Learning and the academic library. *Library Quarterly, 81*(1), 61–82.

O'Clair, K. (2017). Intentionally planning information literacy instruction. In B. West, K. Hoffman, & M. Costello (Eds.), *Creative instructional design: Practical application for libraries*. Chicago: Association of College and Research Libraries (in press).

Scientific method. (2016). In *Oxford English dictionary*. Retrieved from http://www.oed.com/.

Suskie, L. (2009). *Assessing student learning* (2nd ed.). San Francisco: John Wiley & Sons.

Vanderbilt University Center for Teaching. (2001). *Bloom's taxonomy (licensed under Creative Commons 2.0 [no changes made])*. Retrieved from https://www.flickr.com/photos/vandycft/29428436431.

Winkelmes, M. (2016). Self-guided draft checklist for designing a transparent assignment. In *Transparency in learning and teaching project (TILT): Transparent methods*. Available from https://www.unlv.edu/provost/transparency.

CHAPTER 4

Agriculture and Plant Sciences Information Literacy

Livia M. Olsen
Kansas State University, Manhattan, KS, United States

INTRODUCTION

Agriculture happens on every continent except Antarctica. To the general public, agriculture means farming and ranching or similar activities. From the perspective of those working in agriculture or an agricultural researcher, agriculture is much broader and encompasses the foundation on which the modern world is built. The broad subdisciplines within agriculture are plant sciences, animal sciences, and food sciences.

Plant sciences encompass agronomy, horticulture, plant pathology, and related disciplines. Agronomy is what might traditionally be called farming and is concerned with field crops such as wheat, corn, soybeans, and other staple crops that vary regionally. Horticulture is typically associated with fruit and vegetable crops along with ornamental plants. In the United States, one of the biggest crops, various turf grass species, is not grown for food but for beautification (Vinlove & Torla, 1994). Underlying all of the plant sciences, and agriculture generally, is the soil. The plant sciences, in particular, are also concerned with soil health and the micronutrients and microorganisms that reside in the soil. Not only can microorganisms be beneficial, they can also cause disease, and the study of plant pathology is concerned with these fungi, bacteria, and viruses that affect crops. Also relevant across all subdisciplines in agriculture is entomology, because insects can be either beneficial or pestiferous to plants, animals, and food products.

The animal sciences study the production and care of livestock for food, materials such as wool or leather, labor, and recreation. Livestock species include cattle, sheep, goats, horses, poultry, and other less common species. As with staple crops, the importance of a particular livestock species varies, depending on the region. Livestock production systems investigated by agricultural researchers are also variable and include grazing systems on rangelands as well as confinement systems of production. As one would expect, there is a strong link between animal science and veterinary medicine.

Food science research looks at postharvest processing of food and the effects of preharvest management choice on the end product. Food science focuses on meat science, product development, grain science, sensory analysis, and similar topics. Many researchers in this area concentrate on food for human consumption, but others are developing and perfecting products and techniques for feeding animals, also known as feed science. Strong connections exist between human and animal nutrition in food science.

A number of other subjects have connections to agriculture; most notably, agricultural economics. Although agriculture is a science, it is also a business, and the distribution of food throughout the world is vital to human health and well-being. Everyone eats, but not everyone eats well or enough. Agriculture also requires use of vast amounts of land, and food production on that land is affected by climate and weather. Agriculture on a large scale can itself have an effect on climate.

Agriculture has many technological aspects, such as the development of new machinery or techniques for making biofuels. While agriculture is an applied science, agricultural researchers might also delve into topics that increase knowledge across several science disciplines such as genetics and ecology. Sequencing the genome of a crop species and studying the ecological effects of a certain animal on its environment are just two examples. Because agricultural research touches such a variety of disciplines, it can be difficult for first-year students, or even faculty, to know where to seek the information needed to inform their work.

LITERATURE REVIEW

The 2016 *Framework for Information Literacy for Higher Education* (Association of College and Research Libraries [ACRL] Board, 2016) and its six frames have rapidly become the context for any conversation about information literacy in higher education. The literature about agricultural information literacy tends to focus on the sixth frame, Searching as Strategic Exploration. For example, O'Clair (2013) identifies searching Google Scholar and Web of Science as the favored databases of agricultural graduate students. The information-seeking strategies of agricultural producers are also addressed in the literature. Diekmann, Loible, and Batte (2009) identify the motivations and strategies of farmers seeking information, with print media and in-person sources being the most important for producers.

Influencing information-seeking habits of undergraduate agriculture students through library instruction sessions is a common topic for case studies in the field, such as Dinkelman, Aune, and Nonnecke (2010) who describe integrating information literacy and horticulture instruction into an English class. Kesselman and Sherman (2009) discuss embeding information literacy into a class collaborating with local food science businesses. Caminita (2015) presents a case study about embedding a librarian in an agriculture residential college. Debose and Miller (2015) situate information literacy instruction within the lens of undergraduates' first-year-experience curriculum through creating a course about information seeking for freshmen. Rempel and Davidson (2008) identify literature review workshops as being important to providing information literacy instruction to graduate students. Level (2014) presents the most complete analysis of working with faculty to develop and implement information literacy sessions within the curriculum for undergraduate students, including the important information resources and databases for librarians to consult. Much has changed since its publication in 2002, but the book *Using the Agricultural, Environmental, and Food Literature* still provides valuable insights into the agricultural literature (Hutchinson & Greider). Many of the resources mentioned there are now available online.

Not unexpectedly, there is little research on information literacy from the agricultural professor's perspective. One exception is a Delphi study of horticulture professors that identified information literacy skills, such as being able to "identify current trends and topics through current literature and media" and "define a problem and identify resources and alternative solutions" as being important postgraduate skills for horticulture majors (Basinger, McKenney, & Auld, 2009, pp. 456–457). Looking at the habits of agricultural researchers, Kuruppu and Gruber (2006) noted that those researchers "are not aware of…resources and electronic search tools available to them through the library" (p. 615). While that article is more than 10 years old, a forthcoming study across nearly two dozen agriculture institutions expresses the same sentiment from current agricultural researchers (Cooper, 2016).

One group that might also use library services and resources more extensively, but is often forgotten by academic librarians at agricultural research institutions, is staff who work at research farms. Staff are often involved in the everyday aspects of conducting research, which might include such things as caring for animals or planting a crop. Reaching out and making them aware of information resources also benefits student employees who might be working alongside those staff (MacKenzie & Waters, 2014). Most land grant universities also have satellite research stations which, depending on the geographic size of the state, can be located hundreds of miles from the main library. As a result, people working at those farms or research stations often are unaware of the services, including information literacy instruction, provided by librarians on the main campus. In 2007, Davis reported on a project in Florida to reach out to these users to help them gain a better understanding of the library resources and distance services offered. Also included in that project were extension specialists and agents (Davis, 2007). This group is of particular importance when examining the information literacy skills of agricultural producers because these specialists are synthesizing the scholarly literature in preparing their publications and presentations for the public.

DISCIPLINE RESOURCES
Databases
There are several agriculture-focused databases, but due to the expansive nature of agricultural research, some of the most important resources for agricultural researchers are the broad science databases used across the science disciplines. Links to all databases listed here are provided in the Recommended Resources section.

Web of Science Core Collection
One of the most respected literature databases, this resource indexes articles in the sciences, social sciences, arts, and humanities, making it a good starting point for multidisciplinary and interdisciplinary topics. Citations in this database are used to generate journal Impact Factors. Depending on an institution's subscription, up to five citation indexes can be accessed. The most important of these is the Expanded Science Citation Index, with references going back to 1900.

CAB Abstracts
This international database covers all aspects of agriculture, with content in more than 50 languages. In addition to peer-reviewed journal articles, it includes books, cooperative extension publications, patents, and much more. Searching this database is important for animal science researchers applying for institutional animal care and use committee (IACUC) approval prior to conducting research on animals.

PubMed
This premier medical database from the National Library of Medicine also contains information about animal health and veterinary medicine. It is essential for agricultural researchers working in human and animal nutrition, and those researching disease agents, food science, general animal health, and genetics. PubMed is another good source for researchers

seeking IACUC approval. In addition, numerous information products and databases beyond PubMed can be accessed via a drop-down menu on the PubMed homepage.

Google Scholar

This well-known, free database is often a starting place for beginning and advanced researchers alike. While its scope is very broad, it is not clear that researchers use it effectively. Some important discipline resources are not included, so it is important to supplement its use with other databases, if possible.

AGRICOLA

This database from the National Agricultural Library (NAL), US Department of Agriculture (USDA), serves as NAL's library catalog and also searches journal articles, book chapters, and technical reports as well as grant progress reports from USDA-funded research. The public interface can be challenging to use, so some institutions choose to pay to access it through assorted library vendor search platforms rather than using the free NAL interface.

PubAg

This new database from NAL is still a work in progress but is becoming an important search tool for agricultural researchers. It differs from AGRICOLA in that it focuses on the peer-reviewed literature of agriculture.

AgNIC

The Agriculture Network Information Collaboration (AgNIC) consists of libraries at land grant universities and other agricultural research institutions, including the USDA. It has a literature database and keeps track of upcoming agricultural conferences. Some collaborating institutions in AgNIC support their own topic-based websites.

AGRIS

From the Food and Agriculture Organization of the United Nations (FAO), AGRIS contains many millions of records representing agricultural research and technology across the world. It includes scholarly journal articles and other publications, as well as records for data and multimedia works.

Journals

Edited by Hutchinson and Greider (2002), *Using the Agricultural, Environmental, and Food Literature* gives a snapshot of the important resources in agriculture, by subdiscipline, at that time. Many of the main agricultural journals in 2002 were published by society publishers, and this remains true today. Familiarizing new students with the relevant professional societies forms a key part of their understanding of the scholarly communication in any agricultural subdiscipline.

As science, including agricultural science, becomes increasingly interdisciplinary, the major journals for a particular research area are changing. The need to read journals outside of agriculture such as *Geoderma*, *Ecological Modeling*, and *Climatic Change* and broad science journals such as *Science*, *Nature*, and *PLoS One* is becoming more common. Many of the important journals in agriculture live behind expensive pay walls, making access difficult for researchers and students in the developing world. Research4Life is one alternative that provides free or low-cost access to the literature of agriculture for scholars in this group.

Websites

Government agencies conduct research, collect large data sets, and are some of the most copious producers of agricultural information. Agency websites can be excellent sources of information, although it can be challenging for researchers to sort through the many avenues by which the information or data might be obtained. This section focuses on agency websites in the United States; governments in many countries conduct research and collect and disseminate agricultural information and statistics through their own websites. Links to the websites discussed here are provided in the Recommended Resources section.

The USDA collects massive amounts of data on everything from demographics to production to economic indicators in agriculture. Its homepage contains topic-based information drawn from its many subagencies and is a good place to find an overview of a topic. The different types and sheer amount of data available through the different agencies within the USDA are daunting for even an expert. The following list highlights several of the more important USDA products.

- The USDA Census of Agriculture is conducted every five years by the National Agricultural Statistics Service (NASS). This website is organized by topic and state and is searchable. NASS also produces other agricultural information products.
- The USDA Agricultural Research Service (ARS) is the research arm of the USDA and provides information about current USDA initiatives and publications for the public. NAL is a part of ARS.
- The USDA Economic Research Service (ERS) provides economic information as it relates to production, consumption, and income.

- The USDA Natural Resources Conservation Service (NRCS) PLANTS Database contains information about all vascular plants found in the United States, both native and nonnative species.

Numerous other agencies within the United States government are relevant to agriculture. The US Geological Survey, the Environmental Protection Agency, the National Institutes of Health, and the Department of the Interior are just a few.

FAO, in addition to a searchable database, offers extensive information about the state of agriculture and the food system worldwide. There is a strong focus on ending hunger and on sustainability and development.

Weather Data

Accessing reliable weather data is important for agricultural researchers and producers. In the United States, the National Weather Service is the most important provider of up-to-date weather forecasting. Additionally, many states employ a state climatologist or meteorologist who collects and disseminates local weather information. Since precipitation, or lack of it, is often the cause of crop loss and livestock death, there is a National Drought Monitor, housed at the University of Nebraska–Lincoln.

Extension Publications and Presentations

The 1914 Smith-Lever Act created extension services within land grant universities in every US state. These cooperative extension services employ extension specialists who are experts in their field. They publish a wide variety of publications and curricula for agricultural producers and the general public that synthesize the scholarly, peer-reviewed research on a topic for a nonscientific audience. These publications are a good source for students to learn background information on a topic since many current peer-reviewed, scholarly articles are so specific that they can be difficult for new students to understand within the larger, historical scholarly conversation. Although each state's extension service is set up differently, their publications are typically available through the local land grant university's website.

The extension specialists also give presentations around the state and are a great source for fact-based information for agricultural producers. Once extension services were established in the United States, other countries subsequently followed this model of outreach, which is viewed as an "effective way of disseminating information to various agricultural information users, especially farmers" (Aina, 2006).

Books

While journals provide information about the most up-to-date research on a topic, they are usually focused on extremely specific, narrow topics that are not helpful to a researcher seeking general information or a student who wants to learn the basics. For this purpose, books that cover agricultural topics, plant and animal care, anatomy and physiology, genetics, and so forth will be valuable. Other useful books are keys or field guides that help the user identify diseases or a species of plant, animal, or insect. In food science, books provide information about the technical practices used for producing food products or the basic sciences related to nutrition, food chemistry, or biological processes used in manufacturing. Standards established by the US Food and Drug Administration, the National Institute of Standards and Technology, or similar agencies that set quality measures are also important sources of information in food science.

Trade Publications

Trade publications are valuable information sources, particularly for agricultural producers. Some are specific to a sector of agriculture, such as wheat, cattle, or greenhouse production, while others are regional and present a broad swath of information relevant to that geographic area. In addition to the intended audience of producers and industry, trade publications can be important resources for students and researchers who want to stay up-to-date on current trends in technology and the futures market. Perusing the new periodical section in the library of a land grant university reveals a multitude of publications about each livestock species. Like books, these can be good sources of background information on a topic and are written for the general agriculture community rather than researchers, often making them more comprehensible for students beginning their degree programs. These publications can be good sources for generating topics for a research paper. The more science-oriented food industry trade publications even provide citations to research articles.

INFORMATION-SEEKING BEHAVIOR

With so many different resources for information on agriculture, it can be difficult to know where to start. Many researchers, from undergraduate students to faculty, default to searching resources with the widest range of topic coverage, such as Google, Google Scholar, their library's discovery tool, or a handful of larger interdisciplinary databases. In spite of this general tendency, different groups have different information-seeking habits.

Students

New university students tend to rely on search engines and library discovery tools to find all of their sources. They also seek sources recommended by their professors, such as items on reserve at the library or articles and books on a suggested reading list. Because they are not yet familiar with the experts in their subject area, students usually search for information using a topic or keyword rather than a specific researcher's name.

Once students move on to graduate degree programs, their information-seeking habits should change, becoming more sophisticated. While many graduate students know they need better information-seeking skills, they are often trying to function in a highly competitive environment and may not want to acknowledge any lack in this area. These students tend to pick up skills gradually as they learn about searching and glean research tips from their fellow graduate students so that they do not show their ignorance to their professors (Rempel, 2010).

Graduate students learn to use subject-specific databases that focus on scholarly peer-reviewed content and begin to seek out the dissertations and theses of previous graduate students. These publications are important because they provide a comprehensive reference list along with more extensive methodologies than might be found in a typical journal article, which can be very useful to a student working on a similar project. Graduate students also are more likely to use some type of citation management system to organize their search results.

Faculty Researchers

In addition to Google Scholar, faculty use reputable library databases for their searches. Unlike new university students, who are learning their subject area and generally search using keywords, faculty are well-informed subject experts and more likely to search by author for colleagues and other knowledgeable researchers in the field. Reputation is important for faculty in identifying quality research. People in the same disciplines across the world know each other and pay attention to each other's work. Faculty also have developed a standard core of journals relevant to their specialty and regularly scan the table of contents in each new issue. The update function in many databases that emails a user when a new article is published by an author, in a certain journal, or written on a certain topic has been around for a while; however, faculty do not use this functionality as often as one would expect.

Agricultural Producers

There is a body of literature about how and why agricultural producers gather information. Trade publications and personal contacts are important. Producers of agricultural products also look to guidance from land grant universities through extension specialists and agents and from federal government agencies through local offices. In addition to being farmers or ranchers, many are also business owners and may work closely with employees of agricultural companies that develop and market genetics or chemical products. There is very little information about what information literacy skills are expected of employees of agricultural corporations who may not have access to the research literature in the same way those at a university do. A better understanding of what is expected and how employees procure agricultural information is needed.

DISCIPLINE INFORMATION LITERACY INSTRUCTION

Planning information literacy instruction sessions can be a challenge, especially since faculty seeking an instruction session from a librarian are frequently unclear as to expectations about what they want students to learn in a session. Perhaps this is because they do not actually know what they should expect from such a session. Maybe they see it as a find-a-book-on-the-shelf presentation rather than information literacy instruction. A librarian's first job then, is to communicate successfully with the instructor. What can the library offer? What content can be effectively covered in the allotted time span? Is the content best covered in one session or several? Is there an assignment that requires students to use the library? If not, there should be, so have discipline-specific assignments in mind to suggest. This is the time to employ reference interview skills to discover what the faculty and student information needs really are. Give the faculty member a framework for understanding what librarians and libraries can do for their students. Conveniently, the 2016 *Framework for Information Literacy for Higher Education* is a good place to start with this conversation (ACRL Board, 2016). The conversation may go on for many semesters; sometimes getting a foot in the door is a first step to promoting the effectiveness of teaching information literacy long term.

Practical Examples
Movie or Video Clip

In 2016, LaFantasie used a video, *The Story of Stuff*, as a way for students to determine the "credibility and reliability" of information that is presented in an emotional way and with intent to persuade the audience. While the focus of LaFantasie's class was environmental science, the same approach could be taken in a class on agricultural studies because, as with environmental science, this subject also has an abundance of polarizing topics (LaFantasie, 2016). With limited in-class time, clips from a feature-length documentary could be used or a longer video could be assigned for viewing outside of class. During class, students could learn to determine the credibility of the information presented in the video and conduct searches for reliable information.

Interdisciplinary Teaching Teams
One of the most desirable teaching situations is to develop an interdisciplinary team that connects writing, the library, and subject expertise into one class. Dinkelman et al. (2010) discuss an English class taught by an English professor, a librarian, and a horticulture professor in which "skills are taught in the context of the discipline." Kesselman and Sherman (2009) combined academic disciplines, including the library, with an "actual research problem from industry partners" for the students to tackle. Finding ways to deeply integrate information literacy into the curriculum is not always easy, but is one of the best means to show the value of the library to students.

Group Project
A common assignment in upper-level agriculture classes is the group research project, where a small group of students decide on a research topic, conduct the research, and write up the results. Sometimes the final product is prepared as a group with each student submitting a separate literature review; other times, each student submits a separate research paper. These projects usually involve a group presentation covering the topic and research results. Frequently, a professor invites a librarian to speak to such a class because, in past semesters, the literature cited was not of high quality. Students often are not aware of all of the resources available through the library's online system so they default to old search habits, such as using Google and Wikipedia. If the class is in a major where little discipline-based library instruction is given until the students are juniors or seniors, the basics of navigating the library website must be taught along with evaluating sources for quality.

Annotated Bibliography
For students just beginning to conduct research in their major, understanding where their research interests fit within the larger scholarly conversation can be eye opening. Having students produce an annotated bibliography about

their topic is a good way to familiarize them with subject resources. Not only does it help them learn about research already conducted, but the process also helps them connect their own potential research to past efforts by others.

New Literacies Alliance

While not specific to agriculture, the New Literacies Alliance project provides online lessons about information literacy concepts that can be used to gauge student proficiency prior to actually meeting with a class for an instruction session. The initial lessons include the concepts of asking the right questions and understanding the different types of information available; search strategies; access barriers; and the value of information, authority, and citations.

INFORMATION LITERACY INSTRUCTION FOR GRADUATE STUDENTS

Teaching information literacy sessions for graduate students can be particularly problematic. Often, a faculty member's expectation is that in a class time span of 50 or even 15 minutes, the librarian will tell the graduate students about every resource and service that the students might use during their graduate career. Mentioning the many services available might pique interest, but a seminar about literature reviews in their subject area is more useful and relevant to them over time (Rempel, 2010). This is especially true for new graduate students, who likely come to their graduate studies with insufficient information literacy skills, impeding their progress to completing a graduate degree (O'Clair, 2013). Giving these students a foundation on which to base their later research is very valuable. Visiting a group of graduate students more than once is ideal, especially in a seminar format where some depth of instruction is feasible, instead of a one-shot session. Knowing about everything available to them at their library is still important, but much of this kind of information can be covered in a handout or through a librarian-developed resource, such as a LibGuide, rather than through extensive face-to-face instruction.

EMERGING TRENDS AND STAYING UP-TO-DATE

Attending any agricultural lecture at a university, one is likely to hear a question or comment about "feeding the nine billion," a reference to the expected population of Earth in the coming decades. The issue of feeding that many people weighs heavily on the shoulders of agricultural researchers. Combined with climate uncertainties, it is currently the single most important issue in agriculture today. How will the agricultural producers of the world feed that many people and do it sustainably? And what does sustainability even mean? What about resilience? These are major interdisciplinary issues—another big trend in agriculture. Solving problems this large and complex is not just about agricultural research but also crosses into many different disciplines and reflects the trend toward interdisciplinary work throughout the sciences.

Within the library world, there is a trend of library patrons using large interdisciplinary databases for information seeking, leaving the smaller, subject-focused databases at risk of cancellation from lack of use. While a handful of larger databases cover much of the existing scholarly information, the smaller databases still have value, particularly for extremely unique or narrow topics. Convincing younger researchers of the value in searching for information in more than one place is a difficult task. A continuing trend is the move to electronic rather than print journals. In main libraries at land grant institutions, it is difficult to find new issues of scholarly journals on the shelf because more of them are online. The days of faculty coming to the library to browse through the new journals in their field are over, replaced by large interdisciplinary databases with update features for those who choose to use them. It is not clear whether these features are as popular as they should be. Instead, many faculty rely on social networks based on researcher reputation to guide how they seek new information.

For agricultural producers, learning about new trends starts with trade publications and personal contacts (Diekmann et al., 2009). Extension specialists and agents are common sources of trusted information. Many politically or economically conservative producers in rural areas are generally skeptical of scientists, but they do trust their local extension agents to provide reliable science-based information, even on controversial topics (Hibbs et al., 2014).

CONCLUSION

Because agriculture covers such a diverse range of subjects, reflecting the information literacy challenges in each subdiscipline in one book chapter such as this has its limitations. Many questions remain about the information challenges each different group of searchers face and about the information literacy skills each brings to the quest of finding and evaluating sources for a project. Faculty do not always understand the ways in which their library and librarians can help them and their students. Universities want their graduating students to go out into the world and be successful, yet there is little to no research about what types of information literacy skills are expected of those working in the agricultural industry.

Regardless, the 2016 *Framework for Information Literacy for Higher Education* offers a new way forward and aligns more closely with the goals that agriculture faculty have for their students. The Framework offers librarians ways to connect and communicate with faculty and to increase the information literacy skills of students throughout the curriculum. As ever, outreach to lesser-served groups remains important, including service to distance faculty and students, some of whom may be located at research farms; to agricultural staff and their student employees in academic departments; extension agents so they are more aware of library services available to them; and to faculty and students who may not be informed about the possibilities librarians can offer.

RECOMMENDED RESOURCES

AgNIC. Accessed at https://www.agnic.org/.
AGRICOLA. Accessed at https://agricola.nal.usda.gov/.
AGRIS. Accessed at http://agris.fao.org/agris-search/index.do.
CAB Abstracts. Accessed at http://www.cabi.org/publishing-products/online-information-resources/cab-abstracts/.
Food and Agriculture Organization of the United Nations (FAO). Accessed at http://www.fao.org/home/.
Google Scholar. Accessed at https://scholar.google.com/.
National Institutes of Health. Accessed at https://www.nih.gov/.
National Weather Service. Accessed at http://www.weather.gov/.
New Literacies Alliance. Accessed at https://www.lib.k-state.edu/nla.
PubAg. Accessed at https://pubag.nal.usda.gov/pubag/home.xhtml.
PubMed. Accessed at https://www.ncbi.nlm.nih.gov/pubmed.
Research4Life. Accessed at http://www.research4life.org/.
US Department of Agriculture. Accessed at https://www.usda.gov/wps/portal/usda/usdahome.
USDA Agricultural Research Service. Accessed at https://www.ars.usda.gov/.
USDA Census of Agriculture. Accessed at https://www.agcensus.usda.gov/.

USDA Economic Research Service. Accessed at https://www.ers.usda.gov/.

USDA National Agricultural Statistics Service. Accessed at https://www.nass.usda.gov/.

USDA NRCS PLANTS Database. Accessed at https://plants.usda.gov/java/.

US Department of the Interior. Accessed at https://www.doi.gov/.

US Drought Monitor. Accessed at http://droughtmonitor.unl.edu/US.

US Environmental Protection Agency. Accessed at https://www.epa.gov/.

US Geological Survey. Accessed at https://www.usgs.gov/.

Web of Science Core Collection. Accessed at http://wokinfo.com/products_tools/multidisciplinary/webofscience/.

REFERENCES

Aina, L. O. (2006). Information provision to farmers in Africa: the library-extension service linkage. In *Proceedings from World Library and Information Congress: 72nd IFLA General Conference and Council* Seoul, Korea.

Association of College and Research Libraries [ACRL] Board. (2016). *Framework for information literacy for higher education.* Retrieved from http://www.ala.org/acrl/standards/ilframework.

Basinger, A. R., McKenney, C. B., & Auld, D. (2009). Competencies for a United States horticulture undergraduate major: A national Delphi study. *HortTechnology, 19*(2), 452–458.

Caminita, C. M. (2015). Embedding the agriculture librarian in an agriculture residential college: A case study. *Journal of Agricultural & Food Information, 16*(1), 31–42. http://dx.doi.org/10.1080/10496505.2014.984039.

Cooper, D. (2016). *Investigating the needs of agriculture scholars.* Retrieved from http://www.sr.ithaka.org/events/investigating-the-needs-of-agriculture-scholars/.

Davis, V. (2007). Challenges of connecting off-campus agricultural science users with library services. *Journal of Agricultural & Food Information, 8*(2), 39–47.

Debose, K. G., & Miller, R. K. (2015). Stewarding our first-year students into the information ecosystem: A case study. *Journal of Agricultural & Food Information, 16*(2), 123–133. http://dx.doi.org/10.1080/10496505.2015.1013111.

Diekmann, F., Loible, C., & Batte, M. T. (2009). The economics of agricultural information: Factors affecting commercial farmers' information strategies in Ohio. *Review of Agricultural Economics, 31*(4), 853–872. http://dx.doi.org/10.1111/j.1467-9353.2009.01470.x.

Dinkelman, A. L., Aune, J. E., & Nonnecke, G. R. (2010). Using an interdisciplinary approach to teach undergraduates communication and information literacy skills. *Journal of Natural Resources and Life Sciences Education, 39*, 137–144. http://dx.doi.org/10.4195/jnrlse.2010.0005u.

Hibbs, A. C., Kahl, D. W., Pytlik Zillig, L., Champion, B., Abdel-Monem, T., Steffensmeier, T. R., et al. (2014). Agricultural producer perceptions of climate change and climate education needs for the central Great Plains. *Journal of Extension, 52*(3), 3FEA2.

Hutchinson, B. S., & Greider, A. P. (2002). *Using the agricultural, environmental, and food literature.* New York: Marcel Dekker.

Kesselman, M. A., & Sherman, A. (2009). Linking information to real-life problems: An interdisciplinary collaboration of librarians, departments, and food businesses. *Journal of Agricultural & Food Information, 10*(4), 300–318. http://dx.doi.org/10.1080/10496500903245446.

Kuruppu, P. U., & Gruber, A. M. (2006). Understanding the information needs of academic scholars in agricultural and biological sciences. *The Journal of Academic Librarianship, 32*(6), 609–623.

LaFantasie, J. J. (2016). The story of source reliability: Practicing research and evaluation skills using "the story of stuff" video. In L. B. Byrne (Ed.), *Learner-centered teaching activities for environmental and sustainability studies* (pp. 261–266). Switzerland: Springer International Publishing. http://dx.doi.org/10.1007/978-3-319-28543-6_35.

Level, A. V. (2014). Agricultural sciences and natural resources. In P. Ragains (Ed.), *Information literacy instruction that works* (pp. 229–244). Chicago: American Library Association.

MacKenzie, E., & Waters, N. (2014). Library outreach to university farm staff. *Agricultural Information Worldwide, 6,* 114–117.

O'Clair, K. (2013). Preparing graduate students for graduate-level study and research. *Reference Services Review, 41*(2), 336–350. http://dx.doi.org/10.1108/00907321311326255.

Rempel, H. G. (2010). A longitudinal assessment of graduate student research behavior and the impact of attending a library literature review workshop. *College & Research Libraries, 71*(6), 532–547. http://dx.doi.org/10.5860/crl-79.

Rempel, H. G., & Davidson, J. (Winter/Spring 2008). Providing information literacy instruction to graduate students through literature review workshops. *Issues in Science and Technology Librarianship, 53.* http://dx.doi.org/10.5062/F44X55RG.

Vinlove, F. K., & Torla, R. F. (1994). Comparative estimations of US home lawn area. *Journal of Turfgrass Management, 1*(1), 83–97. http://dx.doi.org/10.1300/J099v01n01_07.

CHAPTER 5

Marine and Aquatic Sciences Information Literacy

Sally Taylor
University of British Columbia, Vancouver, BC, Canada

INTRODUCTION

Think for a moment about the importance of water. It covers 71% of the Earth's surface and is critical to life on our planet. The oceans contain about 96.5% of this water with the remaining volume found in ice caps and glaciers, groundwater, ground ice and permafrost, lakes, rivers and swamps, and the atmosphere (US Geological Survey, 2016).

Two-thirds of the world's population lives within 60 km (37 miles) of the coast, and we depend on water for food (e.g., agriculture, fisheries), transportation (e.g., shipping), energy (both nonrenewable sources such as oil and gas, and renewable ones such as tides and wind), biotechnology (e.g., pharmaceuticals), and recreation (e.g., ecotourism) (UN Atlas of the Oceans, 2016). Of course, humans are not the only inhabitants on this planet; nonhuman organisms from tiny microbes to the gigantic blue whale also rely on the aquatic environment for survival but unfortunately are threatened by problems such as climate change and pollution.

Given that water is fundamental to life, it is no wonder that aquatic research is carried out all over the world by academic institutions; federal, regional, and local governments; industry; and nongovernment organizations (NGOs). Although much of the research is interdisciplinary, it is still useful to define the major disciplines before introducing key resources and search strategies.

According to the Association for the Sciences of Limnology and Oceanography (2015), aquatic science is the study of oceanic and freshwater environments. It can be classified as oceanography, the study of the biological, chemical, geological, optical, and physical characteristics of oceans and estuaries; and limnology, the study of these same traits in inland waters (e.g., lakes, rivers, wetlands, etc.). Marine science can be used interchangeably with oceanography but an alternate definition includes marine biology, fisheries, marine resources, and ocean and coastal zone management

(Barnett, 2005). In addition to the broad disciplines, a few other useful terms are marine biology (study of organisms within the marine environment), ichthyology (study of fish), fisheries (industry devoted to catching, processing, and selling fish), aquaculture (rearing of aquatic animals or breeding of aquatic plants for food), phycology (study of seaweeds and other algae), and conservation (preservation, protection, or restoration of the natural environment, ecosystems, vegetation, or wildlife).

Since this book is about information literacy and resources in the life sciences, this chapter will place an emphasis on the biological aspect of marine and aquatic sciences, but it is important to recognize the connection with physical and chemical limnology and oceanography, climate change, conservation, and environmental science.

UNIQUE ASPECTS OF SEARCHING

Before diving into the resources, it is useful to examine the search strategies needed to find information in these fields. As in other disciplines, you will build searches using techniques such as quotation marks for phrases, Boolean logic (AND, OR, NOT), and truncation (★). You will also make use of thesauri to determine subject headings or descriptors; even if only for keyword (vs. subject) searching, it is important to incorporate the controlled vocabulary into the search. So, what are the unique aspects of searching for marine and aquatic information?

First, when teaching students to find information about a biological organism, I emphasize the importance of searching for both the common and scientific names. For example, the species *Oncorhynchus tshawytscha* is known as Chinook salmon, king salmon, Quinnat salmon, spring salmon, and Tyee salmon. The scientific name offers a standard search term that will typically retrieve more results. However, the common name(s) may be appropriate for certain types of literature or where the focus is less on the biology of the organism and more on its utility for humans. Finding common name variants and the scientific name can be accomplished relatively easily, often through a simple Google search.

For those students without a biology background, reviewing the taxonomic classification system used to name and group organisms is a good idea. The scientific name is composed of two parts, the genus and the species within that genus. This naming system, binomial nomenclature, was developed by the Swedish botanist Carl Linnaeus in the 18th century. The genus is always capitalized, and the genus and species name is always italicized, as in the blue whale, *Balaenoptera musculus*. A species may be further divided into subspecies as in the Southern blue whale, *Balaenoptera musculus intermedia*. A person's name and date sometimes follow the scientific name. This identifies the authority or scientist who first published the species name, i.e., *Balaenoptera musculus intermedia* Burmeister, 1871.

Since Linnaeus published his classification system, it has developed into a modern system based on evolutionary relationships, not just structural similarities between organisms. For a recent classification system, see that proposed by Ruggiero et al. (2015). The basic hierarchy is Kingdom–Phylum–Class–Order–Family–Genus–Species but between these can be Super, Sub, and Infra ranks. There are many mnemonics on the Web that can assist with remembering the hierarchy but one common one is: King Philip's Class Ordered the Family Genus to Speak (KFCOFGS). Taxonomy tools for identifying an organism, its scientific name, and the hierarchical classification are listed in the Discipline Resources section.

The second aspect to consider when searching for marine and aquatic science information is its geographical nature. A topic may be connected to a place from the searcher's perspective, but organisms and water bodies do not respect political boundaries. If you imagine the Chinook salmon reproducing in a freshwater stream in British Columbia, Canada, and then making its way to the Pacific Ocean, or the invasive zebra mussel that originated in the Caspian Sea in Eastern Europe but is now found in all five of the Great Lakes in North America (Tans, 2015), you can appreciate that brainstorming search terms will be required not only with regard to taxonomy but also geographical names. A comprehensive strategy should include the names of relevant rivers, lakes, oceans, etc., as well as locations such as islands, countries, regions, and so on. Moreover, when searching for historical information, it is important to know whether political boundaries or location names have changed over time.

The recent proliferation of databases created with geospatial interfaces that allow searching for reference citations or data sets associated with a location on Earth has been especially useful to the aquatic sciences. This exciting development involves indexing an object (e.g., citation, data set, image) with latitude and longitude coordinates to enable searching by points, lines, or polygons on a map. The map links to specific citations and to full text if available. For example, the Coos Bay Georeferenced Bibliography is a publication database maintained in EndNote, but each citation has been georeferenced to the site where the sample was collected (Oregon Institute of Marine Biology Library, 2014; Schmitt & Butler, 2012). Another example is a catalog of data sets hosted by the US government (US General Services Administration,

2017), where it is possible to filter a search to ocean data and then draw a boundary box defined by two latitudes and two longitudes to retrieve data that fall within an area of interest on the map.

Finally, it is worth noting the different publication types and the approaches used for finding information in these sources. All researchers will use journal articles, which are fairly straightforward to find using the commercial databases described below. But they may also require gray literature, which are documents produced outside the commercial realm, typically by academia, government, industry, and NGOs, and not widely distributed. Therefore, it is important to have a good understanding of the landscape of players who produce the information. This is articulated in the ACRL *Framework for Information Literacy for Higher Education*, Searching as Strategic Exploration frame, which states that a learner who is information literate will be able to "identify interested parties, such as scholars, organizations, governments, and industries, who might produce information about a topic and then determine how to access that information" (Association of College and Research Libraries [ACRL] Board, 2016).

When considering a topic or question, teach students to first identify who would conduct the research and disseminate the information. In the marine and aquatic sciences, these sources are most likely to be academic institutions (many of which have research institutes or field stations), government agencies and institutions, intergovernmental organizations, aquaria, museums, companies, and NGOs. In addition to the organizations that publish the information, libraries affiliated with them will likely catalog institutional publications or host them on their websites or in repositories. Librarians at those institutions or organizations can be a font of knowledge for students looking for hard-to-find sources.

DISCIPLINE RESOURCES
Taxonomy and Species Databases

Although reference sources for taxonomy can be purchased, many are now freely available on the Web. They include the following:

Integrated Taxonomic Information System (ITIS)
ITIS is a database of taxonomic information for plants, animals, fungi, and microbes of the world. For each scientific name, ITIS includes the authority (author and date), associated synonyms and common names, taxonomic hierarchy, and source publications. ITIS forms the taxonomic foundation for the Encyclopedia of Life (EOL).
https://www.itis.gov/info.html.

National Center for Biotechnology Information (NCBI) Taxonomy Browser

This is a curated database for all species represented in the NCBI genetic sequence databases, equal to about 10% of described species on Earth. Searching by common name (e.g., blue whale) or scientific name (e.g., *B. musculus*) will find the classification of a species, and hovering your mouse over the taxonomic information will indicate what rank it is (e.g., genus, family, etc.). The record links out to genetic sequence databases, articles in PubMed Central, and specialized tools such as the EOL and the Ocean Biographic Information System.
https://www.ncbi.nlm.nih.gov/taxonomy.

Encyclopedia of Life

Imagine an encyclopedia with a "webpage for every species." That was the goal in 2007 when the EOL was established. It is now an open database of more than one million pages containing text (e.g., species descriptions, ecology, conservation), data (e.g., physical traits, distribution), media (images, video, sound), and maps. Content is incorporated from existing databases and contributed by experts and nonexperts around the world.
http://eol.org/.

AlgaeBase

An aid to taxonomic studies begun in Galway, Ireland, in 1996, AlgaeBase contains information for more than 140,000 species and infraspecies of algae found in marine, aquatic, and terrestrial environments. Entries include taxonomic classification, description, distribution, images, and source publications. There is also a glossary of terms related to algae (e.g., unicellular, red tide, etc.).
http://www.algaebase.org/.

FishBase

FishBase is a global database of more than 30,000 fish species (specifically finfish) and has been available on the Web since 1996. Entries provide information on the taxonomy, distribution, habitat, morphology, ecology, and human uses of individual species. You can search by common or scientific name, but it is also possible to identify species that share characteristics, e.g., multiple species found in the Philippines that are used in aquaculture. There are a number of built-in search tools for finding topical information such as fish sounds, scientific expeditions, and even fish stamps and coins. While FishBase focuses on adult fish, a parallel database called LarvalBase contains information about fish eggs, larvae, and fry.
http://fishbase.org/.

Global Plants
This database is a bit different from other resources in this list in that it includes digitized plant type specimens, contributed by herbaria in more than 300 institutions in 70 countries, which are used to verify nomenclature and act as a record of changes in flora. Global Plants is not a free resource but is offered as a subscription through JSTOR. There are herbaria that provide free access to their specimens, notably the Marine Biological Laboratory of the Woods Hole Oceanographic Institution Library Herbarium and the Cambridge University Herbarium. The latter even includes digitized specimens from Darwin's *Voyage of the Beagle*. http://plants.jstor.org/.

Ocean Biogeographic Information System (OBIS)
OBIS is a global information system focused on marine biodiversity. A species entry may include taxonomy, images, distribution, environmental conditions (e.g., temperature, depth, salinity), occurrences, and data sets. The database includes more than 45 million georeferenced observations for 120,000 marine species from 800 data sources. In addition to searching by species, data sets can be found by country, Marine World Heritage Sites (e.g., Komodo National Park), ABNJ (Areas Beyond National Jurisdiction; e.g., Indian Ocean), and EBSA (Ecologically and Biologically Significant Areas; e.g., Galapagos Islands). OBIS went online in 2002 and became the information system for the 10-year Census of Marine Life initiative. It is now a project of the Intergovernmental Oceanographic Commission of UNESCO's International Oceanographic Data and Information Exchange (IODE) program.
http://www.iobis.org/.

SeaLifeBase
A sister database to FishBase and created in 2006 due to a demand for information on aquatic species other than finfish, SeaLifeBase contains information on the taxonomy, distribution, and ecology for 74,000 marine species. SeaLifeBase links out from a species page to other websites such as Biodiversity Heritage Library and Google Scholar.
http://www.sealifebase.org/.

Encyclopedias

Encyclopedias exist on a wide range of topics. The first two listed here are large databases sold as online subscriptions, although individual volumes of the latter can be purchased outright.

Aquaculture Compendium
Produced by CABI, Aquaculture Compendium is an encyclopedic resource covering all aspects of aquaculture in marine, brackish, and freshwater environments. There are data sheets about fish, mollusks, crustacean, algae, and live feed species as well as diseases, production and environmental systems, and issues in aquaculture. Other features include a database, case studies, images, maps of species' wild and cultured distributions, and maps of disease distributions.
http://www.cabi.org/ac/.

Encyclopedia of Life Support Systems (EOLSS)
An integrated compendium of 21 encyclopedias with peer-reviewed content and arranged thematically, EOLSS includes both major core subjects as well as interdisciplinary articles to "help foster the transdisciplinary context required to fulfill the vision of sustainable development." Sample chapters include *Aquatic habitats in Africa; Biological oceanography;* and *Mangroves of the reef domain: A case study in Belize*. EOLSS was developed under the auspices of UNESCO and EOLSS publishers.
http://www.eolss.net/.

To illustrate the breadth of topics covered, here are just a few examples of encyclopedia titles that can be purchased as ebooks or in print:

Hopley, D. (Ed.). (2011). *Encyclopedia of modern coral reefs: Structure, form and process*. Dordrecht, Netherlands: Springer.

Kornprobst, J. M. (Ed.). (2014). *Encyclopedia of marine natural products* (2nd ed.). Weinheim, Germany: Wiley.

Likens, G. E. (Ed.). (2009). *Encyclopedia of inland waters*. San Diego, CA: Elsevier.

Smith, H. D., Suarez de Vivero, J. L., & Agardy, T. S. (Eds.). (2015). *Routledge handbook of ocean resources and management*. London: Taylor & Francis.

Steele, J. H., Turekian, K. K., & Thorpe, S. A. (Eds.). (2009). *Encyclopedia of ocean sciences* (2nd ed.). San Diego, CA: Elsevier.

Field Guides and Other Identification Guides

A field guide is a book used to identify a species. It typically contains descriptions and illustrations, and often other information such as behavior, habitat, or geographic distribution that will help the reader to distinguish one species from another. One of the best resources devoted to identifying field guides is the International Field Guides database, hosted by the University of Illinois at Urbana–Champaign. It provides a searchable listing, including by classification and biogeographical regions, of field guides for animals and plants worldwide (Schmidt, 2014).

Field guides often have a local or regional relevance, as shown by the following examples:

Davis, R. B. (2016). *Bogs and fens: A guide to the peatland plants of the northeastern United States and adjacent Canada.* Hanover, NH: University Press of New England.

Druehl, L., & Clarkston, B. (2016). *Pacific seaweeds: A guide to common seaweeds on the Pacific west coast.* Madeira Park, BC: Harbour Publishing.

Kells, V. A., Rocha, L. A., & Allen, L. G. (2016). *Field guide to coastal fishes: From Alaska to California.* Baltimore, MD: Johns Hopkins University Press.

Markle, D. F., & Tomelleri, J. R. (2016). *Guide to freshwater fishes of Oregon.* Corvallis, OR: Oregon State University Press.

Reeber, S. (2015). *Waterfowl of North America, Europe, and Asia: An identification guide.* Princeton, NJ: Princeton University Press.

Scientists also use more advanced texts such as atlases, floras and faunas, handbooks, and keys for species identification. Here are a few examples of recently published titles:

Bielanska-Grajner, I., Ejsmont-Karabin, J., & Radwan, S. (2017). *Rotifers (Rotifera): Freshwater fauna of Poland.* Kraków, Poland: Jagiellonian University Press.

Hayward, P. J., & Ryland, J. S. (2017). *Handbook of the marine fauna of North-West Europe.* Oxford: Oxford University Press.

Heesen, H. J. L., Daan, N., & Ellis, J. R. (Eds.). (2015). *Fish atlas of the Celtic Sea, North Sea, and Baltic Sea: Based on international research-vessel surveys.* Wageningen: Wageningen Academic Publishers.

Sim-Smith, C., & Kelly, M. (2015). *Marine fauna of New Zealand: Sponges in the family Geodiidae (Demospongiae: Astrophorina).* Auckland, NZ: National Institute of Water and Atmospheric Research.

Journals

Marine and aquatic science journals are published by large and small commercial publishers, government, universities, and societies. Some are freely available as Open Access journals, but others range in subscription cost from a few hundred to several thousand dollars. It is difficult to estimate how many there are but in a series of publications, Barnett (1986, 1995, 2005) described more than 400 English-language journals and serials in the marine sciences. Core lists of marine journals are also available in multiple editions of *Magazines for Libraries* by Webster and Butler (2006, pp. 706–714, 2008, pp. 648–657, 2010, pp. 582–591, 2012, pp. 569–579). Finally, Journal Citation Reports (2015 ed.), a subscription-based resource that measures how often articles in a journal are cited (i.e., Impact Factor) lists 104 titles in Marine and Aquatic Biology and 19 titles in Limnology.

With so many journals and a finite budget, how does a library decide what to purchase for its researchers? In response to journal cancellations at their institutions, Butler and Webster (2011) questioned the notion of "core journals" and instead recommend analyzing where your researchers are publishing and what they are citing, as well as considering new ways to provide access to journals such as collaborative collection development. When making decisions to renew or cancel a journal, most institutions or libraries look at overall cost, cost per download, Impact Factor, the number of articles authored by their faculty, how often articles are cited from the journal, and local relevance.

Databases

For many researchers in the sciences, the key databases are Google Scholar, Web of Science, or Scopus. Yet, they should be encouraged to make use of specialized databases because of the unique content and powerful search functionality. The databases listed here contain abstracts only, but some will link out to free articles, full text in repositories, or Open URL link resolvers found in many libraries.

Aquatic Sciences and Fisheries Abstracts (ASFA)

ASFA includes international coverage of the world's literature on science, technology, management, and conservation of marine, brackish, and freshwater resources. Dates of coverage vary, and publication types include journal articles, books, conference proceedings, dissertations, films, maps, numerical data, patents, and reports. ASFA comprises five subfiles: (1) ASFA 1: Biological Sciences & Living Resources; (2) ASFA 2: Ocean Technology, Policy & Non-Living Resources; (3) ASFA 3: Aquatic Pollution & Environmental Quality; (4) Aquaculture Abstracts; and (5) Marine Biotechnology Abstracts; as well as complementary access to Oceanic Abstracts. ASFA is produced by the ASFA Partnership, a network of four United Nations agencies and more than 60 international and national partners, each responsible for indexing the aquatic literature in their region. ASFA is published under a cooperative agreement between ProQuest and the Food and Agriculture Organization of the United Nations (FAO).

http://proquest.libguides.com/asfa.

BIOSIS Previews
With international coverage of journal articles, conference proceedings, books, and patents for life sciences and biomedical research, this resource comprises the journal content from Biological Abstracts with supplemental, nonjournal coverage from Biological Abstracts/RRM (Reports, Reviews, Meetings). Coverage dates back to 1926. Published by Thomson Reuters. http://thomsonreuters.com/en/products-services/scholarly-scientific-research/scholarly-search-and-discovery/biosis-previews.html.

CAB Direct
Comprised of two databases—CAB Abstracts and Global Health—CAB Direct includes international coverage of journal articles, proceedings, and books in the applied life sciences such as agriculture (including aquaculture), conservation, global health, and nutrition. Coverage dates back to 1973 with an archive available back to 1900. Published by CABI. https://www.cabdirect.org/.

Environmental Sciences and Pollution Management (ESPM)
Areas of note in this resource include aquatic pollution, bacteriology, ecology, and water resource issues found in journal articles, conference proceedings, books, reports, and government publications in the environmental sciences. ESPM comprises 13 subdatabases, including ASFA 3: Aquatic Pollution & Environmental Quality as well as Water Resources Abstracts. Coverage extends back to 1967. Published by ProQuest. http://proquest.libguides.com/espm.

Fish, Fisheries and Aquatic Diversity Worldwide
Dating back to the 1970s, this database has international coverage of journal articles, conference proceedings, books, reports, and theses in all areas of ichthyology, fisheries, aquatic and marine biology, and aquaculture. It comprises 19 databases, including the ceased database, Aquatic Biology, Aquaculture & Fisheries Resources. It also includes links to FishBase. Published by EBSCO.
https://www.ebscohost.com/academic/fish-fisheries-aquatic-biodiversity-worldwide.

GeoBase
This resource has international coverage of journal articles, conference proceedings, and book chapters in the areas of earth sciences, ecology, geology, human and physical geography, environmental sciences, oceanography, geomechanics, alternative energy sources, pollution, waste management, and nature conservation. Date coverage varies based on subset of records. Published by Elsevier.
https://www.elsevier.com/solutions/engineering-village/content/geobase.

GeoRef
GeoRef indexes journal articles, conference proceedings, books, maps, reports, and theses in the geosciences, including marine geology, oceanography, and hydrology. Published by the American Geosciences Institute and hosted on a variety of platforms, coverage of geology in North America is from 1666 and for the rest of the world it is from 1933. http://www.americangeosciences.org/georef/about-georef-database.

MarinLit
Established in the 1970s at the University of Canterbury, New Zealand, and now published by the Royal Society of Chemistry, MarinLit indexes journal articles related to marine natural products, including new and revised compounds, synthesis, ecology, and biological activities. http://pubs.rsc.org/marinlit/.

Oceanic Abstracts
With coverage going back to 1981, this database indexes journal articles related to the marine and brackish-water environment, including areas such as marine biology, physical oceanography, fisheries, and aquaculture. Published by ProQuest.
http://search.proquest.com/oceanic/productfulldescdetail.

Zoological Record
Zoological Record indexes journal articles, conference proceedings, and book chapters in zoology, including animal behavior, biodiversity, ecology, conservation, taxonomy, and wildlife management, and it serves as the unofficial register of animal names. When the back file is purchased, access is from 1864. Published by Thomson Reuters.
http://thomsonreuters.com/en/products-services/scholarly-scientific-research/scholarly-search-and-discovery/zoological-record.html.

Repositories

For more than a decade, libraries have been digitizing publications and making them available in repositories. Institutional publications, typically available in the past through library exchange programs or subscriptions, are now easily accessible online. Similarly, historical texts found in special collections at select libraries can now be consulted by researchers worldwide. Hundreds of institutional and governmental repositories contain documents relevant to the aquatic sciences. Here are a few key international repositories.

Aquatic Commons
Aquatic Commons is a thematic digital repository focusing on the natural marine, estuarine/brackish, and freshwater environments. It covers the

science, technology, management, and conservation of these environments; the organisms and resources; and the economic, sociological, and legal aspects. The repository is directed by the International Association of Aquatic and Marine Science Libraries and Information Centers (IAMSLIC) and hosted by the UNESCO/IOC Project Office for the International Oceanographic Data and Information Exchange (IODE). http://aquaticcommons.org/.

OceanDocs

OceanDocs is a repository of marine science publications originating from members of the ocean research and observation community around the world. Marine institutions in Africa and Latin America are especially active in contributing documents. OceanDocs is supported by the Intergovernmental Oceanographic Commission (IOC) and hosted by the UNESCO/IOC Project Office for IODE.
http://www.oceandocs.org/.

Biodiversity Heritage Library (BHL) Portal

The BHL Portal is a rich resource of historical literature that includes more than 100,000 titles related to 150 million species names. It is the achievement of an international consortium of natural history and botanical libraries, working in partnership with Internet Archive, to digitize literature held in their collections and make it freely available on the Web. BHL content is incorporated into the EOL.
http://www.biodiversitylibrary.org/.

Government Websites

Websites at all levels of government (e.g., local, regional, national) are useful for providing research publications, library catalogs, repositories (leading to more publications), statistics and data sets, and mapping tools. It is nearly impossible to cover all of the websites of potential use, so instead here is a description of one approach to take when looking for government information in Canada, followed by examples for two other countries.

In Canada, the following federal ministries would be of potential interest: Environment and Climate Change Canada, Fisheries and Oceans Canada, and Natural Resources Canada. Fortunately, the ministries are transitioning to one database for federal government publications (Government of Canada, 2017a) and one Federal Science Library catalog (Government of Canada, 2017b), but until that work is complete, it is also necessary to search the individual ministry sites for additional publications and statistical information. Depending on the topic, websites for the provincial government (e.g., Ministry of Agriculture; Ministry of Environment; Ministry of Forests, Lands and Natural Resource Operations) or even those at the municipal or local levels are good choices.

Always consider who the major players are in specific countries. For example, if you are interested in the United States, key federal agencies are the Environmental Protection Agency (EPA), Geological Survey (USGS), and the National Oceanic and Atmospheric Administration (NOAA), all of which publish a wealth of information. In Kenya, information may be sourced from the Ministry of Environment and Natural Resources, Ministry of Agriculture, Livestock and Fisheries, and the Ministry of Water and Irrigation.

Intergovernmental Websites

Since water bodies and species cross national boundaries, a number of intergovernmental organizations have been created to inform management of these resources. Similar to government sites, they are a good source of publications, statistics and data, and maps.

Food and Agriculture Organization
The three main goals of FAO are eradication of hunger, food insecurity, and malnutrition; elimination of poverty and the driving forward of economic and social progress for all; and sustainable management and utilization of natural resources, including land, water, air, climate, and genetic resources for the benefit of present and future generations. FAO has a publications repository, statistics tools for fisheries and water, and a catalog of the holdings at the David Lubin Memorial Library, renowned for its resources on food, agriculture, and international development.
http://www.fao.org/home/en/.
http://www.fao.org/library/libraryhome/en/.

Inter-American Tropical Tuna Commission (IATTC)
Responsible for the conservation and management of tunas and other marine resources in the eastern Pacific Ocean, IATTC publishes monthly catch statistics and maintains a registry of vessels. Publications include its Bulletin, Stock Assessment Reports, Special Reports, and Data Reports.
https://www.iattc.org/HomeENG.htm.

International Commission for the Conservation of Atlantic Tunas (ICCAT)
Monitoring conservation of tunas and tunalike species in the Atlantic Ocean and adjacent seas, ICCAT compiles statistics for species in the Atlantic Ocean and offers other resources on its site, including a record

of vessels, a record of bluefin tuna farming facilities, and a list of tagging programs. Series based on its work include Collective Volume of Scientific Papers, Statistical Bulletin, and Biennial Reports.
https://www.iccat.int/en/.

International Council for the Exploration of the Sea (ICES)
This global scientific organization facilitates research between its 20 member countries and advises decision makers on the sustainable use of the marine environment and ecosystems. ICES publications include its subscription-based *ICES Journal of Marine Science*, as well as Cooperative Research Reports synthesizing research conducted by expert groups, and Conference and Meeting documents. It also maintains a data portal covering such themes as biodiversity (seals and seabirds), fish egg and larvae (from ichthyoplankton surveys), and underwater noise.
http://www.ices.dk.

International Joint Commission (IJC)
The IJC is an example of an organization formed by two countries (Canada and the United States) to prevent and resolve disputes and to pursue the common good as it pertains to lake and river systems along their border. Guided by the Boundary Waters Treaty of 1909, its jurisdiction applies to a range of water uses including drinking water, shipping, hydroelectric power generation, agriculture, industry, and fishing. Its website includes technical reports, administrative documentation, and interactive maps.
http://www.ijc.org/.

International Pacific Halibut Commission (IPHC)
Established by Canada and the United States in 1923 to research and manage the stocks of the Pacific halibut, IPHC conducts standardized stock assessment fishing surveys and basic halibut biology projects. Annual survey data, scientific reports, technical reports, and information bulletins are found on the website.
http://www.iphc.int/.

International Whaling Commission (IWC)
IWC was established under the International Convention for the Regulation of Whaling in 1946 as the global organization charged with the conservation of whales and the management of whaling. Although a moratorium on commercial whaling was introduced in 1986, IWC continues to set catch limits for aboriginal subsistence whaling. It also works to address issues such as entanglement, ship strike, marine debris,

and climate change. IWC maintains a library of full-text publications and publishes the *Journal of Cetacean Research and Management*. Its website links to related research organizations and agencies in both Canada and the United States, as well as worldwide, including the United Nations Oceans & Laws of the Sea.
https://iwc.int/home.
Northwest Atlantic Fisheries Organization (NAFO)
NAFO is a multinational, intergovernmental body whose main objective is to aid in the optimum utilization, management, and conservation of fishery resources (excluding salmon, tunas/marlins, whales, and sedentary species) in the NAFO Convention Area. Charged with making fisheries research pertaining to the Northwest Atlantic available to the scientific community, its website provides access to catch statistics, the peer-reviewed *Journal of the Northwest Atlantic Fishery Science,* and other publications.
https://www.nafo.int/.
Pacific Salmon Commission (PSC)
PSC was formed to implement goals set out in the Pacific Salmon Treaty of 1985 between Canada and the United States to conserve and manage the shared key resource of Pacific salmon. PSC maintains a staffed research library, and its website links to fisheries maps, technical and annual reports, and many other publications including those of its predecessor, the International Pacific Salmon Fisheries Commission.
http://www.psc.org/.

Data and Map Sources

Many of the websites described above contain data or maps; this section highlights resources where these are the main focus.
AQUASTAT
AQUASTAT is FAO's global water information system, developed by its Land and Water Division. It provides reports, data sets, summary tables, maps, country profiles, and river basin profiles related to water resources, water uses, irrigation and drainage, dams, and climate.
http://www.fao.org/nr/water/aquastat/main/index.stm.
Dryad
This digital repository is an Open Access curated resource of data sets linked to specific scientific publications that can be searched by keyword and then filtered by author, subject, or journal.
http://datadryad.org/.

FAO Fisheries and Aquaculture Department

To promote responsible aquaculture and fisheries, FAO analyzes and disseminates global statistics on production (both capture and aquaculture), tuna catches, number of fishers, consumption of fish and fishery products, and fishery commodities and trade. It offers the data in three ways: a stand-alone application called FishStatJ for experts and scientists; online query panels for advanced users to extract customized reports; and the FAO *Yearbook of Fishery and Aquaculture Statistics*, an annual summary with detailed tables.

http://www.fao.org/fishery/statistics/en.

Global Biodiversity Information Facility (GBIF)

GBIF is an international, open data initiative funded by governments and organizations around the world to allow anyone, from student to policy maker to researcher, to access and use data and information about all types of life on Earth. Underway in 2001, this resource can be explored in a myriad of ways, including species, country, and data publisher; results are georeferenced on a world map.

http://www.gbif.org/.

PANGAEA

PANGAEA is a repository of georeferenced data for earth and environmental sciences. While most data are freely available, some data sets for ongoing projects may be under embargo; if so, a description and the investigator's contact information are provided. PANGAEA can be searched by keyword and filtered by criteria such as author, year, theme (e.g., oceans), device, and location (e.g., Weddell Sea). Data sets can also be viewed using Google Maps or Google Earth.

https://www.pangaea.de/.

ReefBase

A global information repository for coral reefs, ReefBase comprises three main tools: a Global Database that provides data and information on the location, threats, monitoring, and management of coral reefs in more than 120 countries and territories; the ReefBase Online Library of nearly 30,000 publications; and ReefGIS, offering access to maps on themes such as coral disease, reef bleaching, and marine protected areas.

http://www.reefbase.org/.

Sea Around Us

A research initiative at the University of British Columbia to assess the impact of fisheries on the marine ecosystems of the world, this source includes catch data that can be searched by country or taxonomy. Results can be displayed at spatial scales with ecological and policy relevance, such

as by Exclusive Economic Zones, high seas, or large marine ecosystems. Data are also available for fisheries economics, biodiversity, and mariculture. Graphed data are reconstructed based on official reported data (mainly from FAO) and reconstructed estimates of unreported data from the literature and consultations with local experts.
http://www.seaaroundus.org/.
Water Resources of the United States
This site serves as an example of the kind of data related to water that governments may disseminate. The US Geological Survey provides past and present data for such topics as streamflow, floods, droughts, surface water quality (e.g., temperature, pH), and groundwater.
https://www2.usgs.gov/water/.
World Database of Protected Areas (WDPA)
This global database of marine and terrestrial protected areas is presented through a mapping interface. Searches can be conducted for protected areas by individual names or via a map that identifies protected areas in a region. WDPA is a joint project between the United Nations Environment Programme (UNEP) and the International Union for Conservation of Nature (IUCN). It is managed by UNEP World Conservation Monitoring Centre (UNEP-WCMC).
https://www.protectedplanet.net/.

Citizen Science

The world is a big place. What if you could harness the energy of amateur naturalists to gather observations that scientists could use? Often using mobile apps, citizen scientists are doing just that. You may be familiar with eBird, but there is a raft of others, including many for the aquatic world. Here are a few global examples.
iNaturalist
iNaturalist covers the terrestrial and aquatic world. Participants record an observation by taking a photograph, share it with other naturalists, and discuss the findings to crowdsource a correct identification. iNaturalist shares its data with other data repositories like the GBIF described above.
http://www.inaturalist.org/.
iSeahorse
The goal of iSeahorse is to better understand seahorse behavior, species ranges, and threats to improve conservation around the world. Participants can submit a photo, description, coordinates, and even guess the species based on identification guides available on the website.
http://www.iseahorse.org/?q=home.

Reef Life Survey (RLS)
Recreational divers participate in the RLS to collect data at scales impossible for researchers to cover. Divers use standardized visual census methods to record fish and invertebrate species seen along underwater transects. In addition to being experienced and capable, divers have to be dedicated enough to identify species and enter the data after their dives. http://reeflifesurvey.com/.

Secchi Disk: The Global Seafarer Study of the Phytoplankton
Phytoplankton produce 50% of the world's oxygen and are a key part of the aquatic food web, but scientists have observed a population decline in recent years. A Secchi disk provides a simple way to measure water clarity, and when measured away from estuaries and shallows, it is an indication of the amount of phytoplankton in the water column. Seafarers and boaters are asked to lower the disk vertically from a stationary boat and record their GPS coordinates and the Secchi depth, the point at which the disk disappears from sight. The disk and the results are available on the website.
http://www.secchidisk.org/.

DISCIPLINE INFORMATION LITERACY INSTRUCTION
Undergraduate Students

My main goals for undergraduate students are to make sure they know the resources available to them and how to conduct a basic search effectively. When demonstrating a database, I develop a research question that will enable me to illustrate the concepts of phrase searching, Boolean logic (AND, OR, NOT), and truncation, as well as how to search for the scientific name. In an upper-division class and certainly in a one-on-one consultation, it is appropriate to explain and demonstrate how using subject headings/descriptors can be an effective way to increase the relevancy of results. Depending on the database, it may be useful to discuss key features such as sorting, filtering/limiting, and saving/exporting records. The latter is particularly helpful if citation management software is available to the students.

As an example of the process, a sample research question that is drawn from real life and relates to the students' coursework might be: Migration of Chinook salmon from the river to the ocean is being impeded by hydroelectric dams along the route. What strategies are being tested to increase the survival rate of the smolts? I query the class for the two or three main ideas in the question, teaching them to identify the individual components and terms. Once the question is understood, we can examine how to choose an appropriate database, one that is likely to hold applicable content and what its authority is based on. This is often a good time for discussion about the differences between subscription databases, Google, and Google Scholar. As the search strategy is explained and developed within the structure of the chosen database, I demonstrate each step of the way through searching the database, showing the kind and number of retrieved results and asking the class what they would do next. This generally allows for instruction of all of the basic concepts. For this research question example, developing the search strategy might look like this:

- Identify main topic. Introduce phrase searching.
  "Chinook salmon"
- Determine and include the scientific name. Demonstrate how the Boolean operator OR means more.
  "Chinook salmon" OR "oncorhynchus tshawytscha"
- Include the second concept. Is there a synonym that should be used? Introduce the use of AND.
  ("Chinook salmon" OR "oncorhynchus tshawytscha") AND (dam or dams)
- Include the third concept. Introduce truncation.
  ("Chinook salmon" OR "oncorhynchus tshawytscha") AND (dam or dams) AND surviv*

A very basic introduction to Boolean operators for first- or second-year students in a large lecture hall is an exercise that my colleagues and I call "Stand-up Boolean." The students find it fun, and it can give them a visceral understanding of Boolean searching, or at least a good laugh. It works like this:

Please stand up if you had salad for lunch.
Please stand up if you had salad OR a sandwich for lunch.
Please stand up if you had salad OR a sandwich OR sushi for lunch.
Please stand up if you had salad OR a sandwich OR sushi for lunch AND a juice.

Then we debrief. When OR is used, more and more students stand up. When AND is included, only a few students are left standing.

The online format of information somewhat masks the purpose and quality of a resource, so it is important for undergraduate students to understand what they are looking at when they find an article or book or website. When teaching a conservation course, it works well to pick a hot topic or

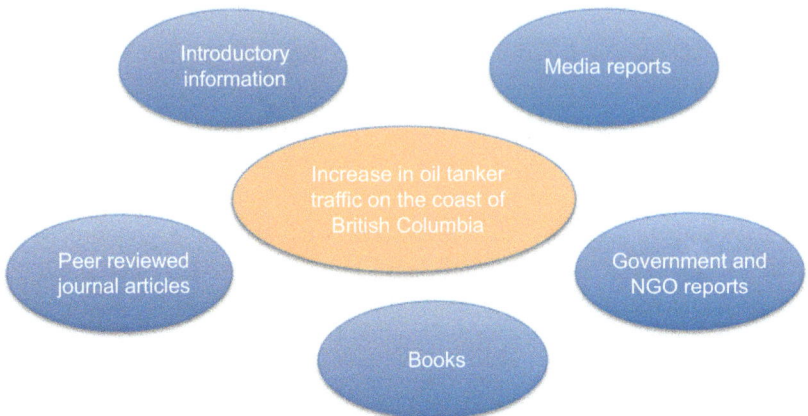

Figure 5.1 Example of topic and associated literature sources. *NGO*, nongovernment organization.

one that is more controversial (e.g., marine protected areas, invasive species, ocean acidification), provide a list of the types of information (e.g., media, books, journal articles, reports), and have the students discuss in pairs when they would use that type of information and how they would look for it. I would project this diagram on the screen (Fig. 5.1).

The students contribute what they know, which can then be reinforced by repeating or paraphrasing, or clarifying or correcting if necessary. I also take this opportunity to explain peer review and distinguish it from review articles. This exercise works particularly well because it engages students who already know something about evaluating information and can share it with their peers. It also leads nicely into the searching component of the class.

Graduate Students

Sometimes graduate students may not know much more than undergraduate students with respect to searching. Perhaps they made it through their undergraduate degrees using only Google Scholar, or they have returned to university after time away, or they may have studied previously at an institution without specialized databases. One big difference is that they usually are already familiar with their area of research, so I typically ask them for an example of a research question or information need. Using the student's example is also more authentic, slows down the demonstration, and illustrates that searching is an iterative process.

Graduate students, of course, have greater information needs than undergraduates because their research is more comprehensive. The specialized databases that index gray literature are especially beneficial to them. Graduate students are the scholars of the future, so incorporating techniques for identifying key authors or journals in their fields and introducing them to the issues of scholarly publishing, Open Access, and research data management are valuable skills for them to learn. One-on-one or small group consultations can be the most rewarding because questions can be much more in-depth. An example of a more focused project used in the past is one where graduate students and other research assistants on the University of British Columbia's Sea Around Us project had to search for catch statistics beyond the reported FAO data. Strategies included searching for scientific publications using ASFA, tracking down government websites (often in other languages), and being cognizant of changing names of government bodies and even countries.

STAYING UP-TO-DATE

If you are passionate about the aquatic environment, it is relatively easy to stay current by following news from the media and aquatic organizations on Twitter or whatever your favorite social media tool may be. For information about new resources and trends in information management, library associations can be invaluable.

In this field, we are fortunate to have an entire library association devoted to the topic. The IAMSLIC brings together more than 300 members from around the world in academia, government, industry, and NGOs to explore ideas and issues of mutual concern. IAMSLIC began under the name East Coast Marine Science Librarians in 1975 with a meeting of 23 librarians from the United States, Canada, and Bermuda. Since then it has expanded geographically and encompasses all aquatic sciences, not just marine. IAMSLIC has six regional groups, so members can forge connections closer to home. Like other associations, IAMSLIC has a blog and an email list, and hosts an annual conference with published proceedings.

Over the years, my friends and colleagues in IAMSLIC have taught me a lot about key areas such as digitization, repositories, Open Access, data management, and preservation. One of the things I appreciate most is the international perspective of the members, illustrated by the four resources below that were presented at recent IAMSLIC conferences.

Agriculture and Environmental Data Archive
A project managed by the UK Freshwater Biological Association in partnership with the Centre for e-Research at King's College London, this archive contains data sets, gray literature, images, video, and a variety of other information.
http://www.environmentdata.org/.

Gaia Antarctic Digital Repository
The purpose of this resource is to develop a single platform to collect, preserve, and disseminate information about Antarctic and polar topics. The repository is a partnership of the Universidad de Magallanes in Chile with academic and research institutions contributing content.
http://antarticarepositorio.umag.cl/.

Oregon Estuarine Invertebrates: Rudys' Illustrated Guide to Common Species (3rd ed.)
While most field guides are only available for purchase in print or occasionally as an ebook, *Oregon Estuarine Invertebrates* is freely available from the University of Oregon's institutional repository.
http://researchguides.uoregon.edu/oei.

SPC Coastal and Oceanic Fisheries Digital Library
The Secretariat of the Pacific Community (SPC) is the principal scientific and technical organization in the Pacific region owned and governed by 26 country and territory members. The Digital Library helps with the dissemination of fisheries and aquaculture-related documents produced by, for, or in collaboration with SPC to fisheries managers throughout the Pacific.
https://www.spc.int/DigitalLibrary/FAME.

FINAL THOUGHTS

I hope the selection of resources and examples has helped you to appreciate the importance of marine and aquatic sciences and the breadth of research that is available to students and researchers alike. While many resources have been covered here, many more are out there to be discovered. It is impossible to know them all, so remember to look at the context of the question (e.g., species, location, issue) and consider which organization(s) would be compelled to collect data and publish on the topic. Also check other library websites to see what resources they recommend, and if you join IAMSLIC, you will have an international network of colleagues with whom to consult.

REFERENCES

Association for the Sciences of Limnology and Oceanography. (2015). *Careers in the aquatic sciences*. Retrieved from http://www.aslo.org/information/aquaticcareer.html.

Association of College and Research Libraries [ACRL] Board. (2016). *Framework for information literacy for higher education*. Retrieved from http://www.ala.org/acrl/standards/ilframework.

Barnett, J. B. (1986). *Marine science journals and serials: An analytical guide*. Westport, CT: Greenwood Press.

Barnett, J. B. (1995). Marine science journals and serials: An update. *Science & Technology Libraries, 15*(1), 3–22. http://dx.doi.org/10.1300/J122v15n01_02.

Barnett, J. B. (2005). Marine science journals and serials. *Science & Technology Libraries, 25*(4), 87–102. http://dx.doi.org/10.1300/J122v25n04_07.

Butler, B., & Webster, J. (2011). Core journals: Fact or fiction? In *36th IAMSLIC Conference* Mar del Plata, Argentina.

Government of Canada. (2017a). *Federal science library*. Retrieved from http://cat.cisti-icist.nrc-cnrc.gc.ca/search.

Government of Canada. (2017b). *Publications catalogue*. Retrieved from http://www.publications.gc.ca/site/eng/home.html.

Oregon Institute of Marine Biology Library. (2014). *Coos Bay georeferenced bibliography*. Retrieved from https://www.journalmap.org/oregon-institute-of-marine-biology-library/coos-bay-georeferenced-bibliography.

Ruggiero, M. A., Gordon, D. P., Orrell, T. M., Bailly, N., Bourgoin, T., Brusca, R. C., et al. (2015). A higher level classification of all living organisms. *PLoS One, 10*(4), e0119248. http://dx.doi.org/10.1371/journal.pone.0119248.

Schmidt, D. (2014). *International field guides*. Retrieved from http://www.library.illinois.edu/bix/fieldguides/.

Schmitt, J., & Butler, B. (2012). Creating a geo-referenced bibliography with Google Earth and GeoCommons: The Coos Bay bibliography. *Issues in Science and Technology Librarianship, 71*. http://dx.doi.org/10.5062/F43J39XH.

Tans, E. D. (2015). Reviews of science for science librarians: The Great Lakes fishery – current status, human impacts, and literature. *Science & Technology Libraries, 34*(1), 43–66. http://dx.doi.org/10.1080/0194262X.2015.1005275.

UN Atlas of the Oceans. (2016). *Leaflet*. Retrieved from http://www.oceansatlas.org/.

US General Services Administration. (2017). *Data.gov data catalog*. Retrieved from https://catalog.data.gov/dataset.

US Geological Survey. (2016). *How much water is there on, in, and above the earth?*. Retrieved from https://water.usgs.gov/edu/earthhowmuch.html.

Webster, J., & Butler, B. (2006). Marine science and technology. In C. LaGuardia, & W. A. Katz (Eds.), *Magazines for libraries* (15th ed.) (pp. 706–714). New York: Bowker.

Webster, J., & Butler, B. (2008). Marine science and technology. In C. LaGuardia, & W. A. Katz (Eds.), *Magazines for libraries* (17th ed.) (pp. 648–657). New York: Bowker.

Webster, J., & Butler, B. (2010). Marine science and technology. In C. LaGuardia, & W. A. Katz (Eds.), *Magazines for libraries* (19th ed.) (pp. 582–591). New York: Bowker.

Webster, J., & Butler, B. (2012). Marine science and technology. In C. LaGuardia, & B. Katz (Eds.), *Magazines for libraries* (21st ed.) (pp. 569–579). New York: Bowker.

CHAPTER 6

Polar (Arctic and Antarctic) Sciences Information Literacy

Sandy Campbell[1], Jessica Thorlakson[1], Julianna E. Braund-Allen[2]
[1]University of Alberta, Edmonton, AB, Canada; [2]University of Alaska Anchorage, Anchorage, AK, United States

WHY A CHAPTER ON POLAR SCIENCES INFORMATION LITERACY?

Other chapters in this book describe information resources for various disciplines within the life sciences. For an undergraduate student writing a basic research paper that might require 10 peer-reviewed works on the natural history of a particular species, standard resources such as Zoological Record or BIOSIS databases and other materials most likely found in a local library collection will suffice. Information literacy training for a student at that early point in learning might well be straightforward and consist of how to find a book in the library's catalog and how to use a subject database or Google Scholar to find a journal article. However, if the project extends beyond this level and involves polar information, one quickly finds that standard, peer-reviewed polar literature is not like the tip of an iceberg, where 90% is below the surface but completely predictable in volume and location. Instead, the peer-reviewed literature in polar studies is more like ice on a lake, a thin visible veneer that covers the surface, while below, the information may be deep or shallow, may be known to only a few, and may be sparse or even unique to one location; it may exist in unfamiliar organizational systems, and it may be hidden from our normal ways of searching. Efficient navigation through polar literature sources has to be learned, and it is best learned from a guide—hence this chapter.

DEFINING INFORMATION LITERACY IN THE POLAR CONTEXT

In 2000, the Association of College and Research Libraries (ACRL) supplied a definition of information literacy for higher education that, while recently replaced, remains useful for the literature of the Arctic and Antarctic. It contained five basic points: knowing that you need information,

Agriculture to Zoology
ISBN 978-0-08-100664-1
http://dx.doi.org/10.1016/B978-0-08-100664-1.00006-5

Copyright © 2017 Daria O. Carle, Julianna E. Braund-Allen and Jodee L. Kuden. Published by Elsevier Limited. All rights reserved identifying the information you need, accessing the information you need, evaluating the information you find, and applying the information to your needs (ACRL, 2000). Librarians have interpreted this definition for their own environment (Fabbi, 2012; Scaramozzino, 2008). The definition applies to the satisfaction of every kind of information-seeking activity, encompassing both library and Internet-based research (Garcia, 2014). It equally applies to someone seeking information only available from an Indigenous elder (Campbell, 2008). Someone who is literate in polar information will be liberal in identifying their sources, creative in accessing them, persistent in acquiring the content, and flexible in the evaluation and application of the information.

In 2006, ACRL's Science and Technology Section Task Force on Information Literacy for Science and Technology developed competency standards specific to science, engineering, and technology. Here, information literacy was presented as a set of abilities that included being able to identify, find, and evaluate information across a wide variety of information sources; to be able to revise strategies as needed for finding information; and to use the information in an ethical and legal manner. The five standards included 25 performance indicators as well as multiple assessment outcomes, reflected a rapidly changing and increasingly complex information and research environment, and emphasized the ensuing importance of student competency. The Task Force also contextualized the disciplines of science, engineering, and technology, highlighting the unique challenges posed by gray literature, interdisciplinary research, and information formats. The latter can vary widely from technical reports and data sets to geographical information systems and often require knowledge of specialized software to access and use (ALA/ACRL/STS Task Force on Information Literacy for Science and Technology, 2006). These standards still apply and are particularly relevant to polar information literacy. Students seeking polar information must consider the interrelated questions of whether the information exists, who might have produced it, and how it might have been made available. They need the skill sets to evaluate the sources and possible influences on the sources, as well as the ability to be flexible in their search strategies and to make use of the wide array of formats in which polar information appears.

In 2016, ACRL formally adopted a new *Framework for Information Literacy for Higher Education,* which replaced the older 2000 standards. The Framework "envisions information literacy as extending the arc of learning throughout students' academic careers and as converging with other academic and social learning goals," offering an expanded definition of information literacy that "emphasize[s] dynamism, flexibility, individual growth, and community learning" (ACRL, 2016). The Framework stresses teaching critical and self-reflective thinking skills to give students a holistic context for lifelong information literacy in the real-world digital environment. ACRL considers "information literacy as an educational reform movement," and the new Framework "depends on [the] core ideas of metaliteracy" (ACRL, 2016). The Framework quotes Mackey and Jacobson (2014), who write that metaliteracy:

> Builds on decades of information literacy theory and practice while recognizing the knowledge required for an expansive and interactive information environment. Metaliteracy expands the scope of traditional information skills…to include the collaborative production and sharing of information in participatory digital environments (collaborate, participate, produce, and share)…[It] requires an ongoing adaptation to emerging technologies and an understanding of the critical thinking and reflection required to engage in these spaces as producers, collaborators, and distributors (p. 1).

The 2016 Framework underscores the need to engage students in developing critical thinking skills from the onset. However, except when required by their discipline faculty, many students may not choose to interact with librarians, believing that on their own they can find, sift through, and understand all they need on the Web through a search engine. As librarians, we know this is not the case, and it is especially not the case when it comes to polar information.

What Do We Mean by Polar?

When teaching students to know where to look for Arctic and Antarctic information, the first important concept to introduce is the geography of the poles, both north and south, which can be demonstrated with a globe or a map. It may seem silly to say that students need to learn where the Arctic and Antarctic are physically located, but many people have never looked at a globe from the top and bottom, seen a polar projection map, or paid particular attention to the Polar Regions. Understanding which countries physically surround the Arctic Ocean and which countries are involved in the Antarctic, as well as their social, political, and economic relationships, is critical to comprehending the generation of information about these regions. For biological information, attention needs to focus on both the marine and land areas of the Polar Regions. For an example of an information literacy assignment developed to teach the geography of the Polar Regions, see Polar Information Activity 1 in this chapter's Polar Information Literacy Instruction section.

What We Know About Polar Information Sources

While having many similarities in being cold, ice covered, dark part of the year, remote, and sparsely populated as well as possessing large marine components, the individual characteristics of the Arctic and Antarctic are singular enough to make information about the two areas distinct. Moreover, we know that producing and recording information is inherently expensive even for areas that are warmer and more easily accessible. When looking for information, start by considering who would have paid to have had polar information collected, recorded, and/or published. For the Antarctic, the initial producers of information were explorers and mariners. Since then, most of the information produced has been generated by academics or by governments and international organizations with an interest in the Antarctic.

The Arctic's publication history differs from that of Antarctica. The Circumpolar North refers to the northern lands, both Arctic and Subarctic, of the world's eight northernmost countries. Often referred to as the Arctic Eight, these countries are Canada, Denmark (including Greenland and the Faroe Islands), Finland, Iceland, Norway, Russia, Sweden, and the United States (Alaska). Each of these countries has developed its own libraries, repositories, and organizational systems. The Arctic, like the Antarctic, has exploration and expedition histories, but many different groups, from governments to private citizens, publish materials on and about the Arctic. Consequently, sources of information about the Polar North are much more complex and varied than those of the Polar South.

Additionally, vast polar areas with little human population exist, and there are many polar subjects about which little to nothing is published in scholarly literature. In these cases, researchers must turn to less formal sources, for example, to newspapers and newsletters; to primary sources, such as a missionary's journal describing caribou populations; to data collected by others, such as that amassed during the International Polar Years (IPYs); or to collecting the data themselves, as in asking an Inuit elder what kind of herb is best for treating headaches.

WHERE TO LOOK FOR POLAR INFORMATION

Specialized libraries and organizational systems for polar information developed many years ago due to the complexity and difficulty involved in identifying and locating Arctic information sources, the range of ways in which polar information is used, and the variety of geopolitical interests present in the Arctic and Antarctic. Many of the libraries that were established were geospecific, reflecting the need to locate information by place. The Universal Decimal Classification (UDC) scheme that was under development in Europe around the end of the 19th century proved to be especially well suited for organizing materials by location as well as by subject (UDC Consortium, 2017). The Library of Congress (LC) Classification used by most North American academic libraries does not allow for collections to be organized by geographic location. LC Classification assumes that researchers approach questions first by subject, and then geographically, which disperses polar materials throughout subject collections. Geographic subheadings are often not even assigned in LC unless the geographic component is highly obvious. Because the polar physical environments exert a strong influence on everything biological, the geographic aspects and location of any given research are paramount. As a result, specific polar collections developed that used UDC or other specialized schemes for colocating polar materials and providing good access where LC Classification fails.

Antarctic

Countries with Antarctic interests, such as the original 12 signatories of the 1959 Antarctic Treaty (Argentina, Australia, Belgium, Chile, France, Japan, New Zealand, Norway, South Africa, the Soviet Union, the United Kingdom, and the United States), created national organizations that function today as the primary conduits through which Antarctic interests are managed. The original 12 countries to the treaty were those whose researchers had been involved in Antarctica during the 1957–58 International Geophysical Year. Today there are more than 40 additional signatories (Secretariat of the Antarctic Treaty, 2017) that have also created or joined national or other organizations to manage their interests. Many of these entities developed library and museum collections to make Antarctic information available. Examples include the British Antarctic Survey Library in Cambridge, United Kingdom; the Australian Antarctic Division's library in Hobart, Tasmania; and the library of the Programma Nazionale Di Ricerche in Antartide (PNRA) in Rome.

Arctic

Similarly, Arctic-focused collections, largely housed in academic institutions, were developed over time to create research efficiency by colocating hard-to-find Arctic materials. Examples of these collections include the Arctic Center at the University of Lapland in Rovaniemi, Finland; the Boreal

Institute Library (later the Canadian Circumpolar Library) at the University of Alberta in Edmonton, Canada; the Rasmuson Library, Alaska and Polar Regions Collections at the University of Alaska Fairbanks; and the Alaska Resources Library and Information Services colocated with the UAA/APU Consortium Library at the University of Alaska Anchorage.

A LISTING OF POLAR INFORMATION RESOURCES

This section compiles the primary resources for finding polar information and for using as a basis to build polar information literacy. Many of the best life sciences resources are priced products and not all libraries will have access to them. For this reason, we have attempted to point to Open Access (OA) or shared resources so that the examples will work for all readers. However, some classic or defining works are only available as priced products and must be mentioned because of their importance to the field. While there has been much progress in the development of digital resources, many classic works remain only available in print format, making the resources listed in this chapter a mix of digital and print.

Polar Libraries

Teaching students to locate and search polar library collections is a good starting point for a polar information literacy program. Searching library holdings may identify unusual resources or confirm that nothing is available on the student's subject. Fortunately, identifying libraries with collections to search is not difficult. In the early 1960s, many polar libraries came together to create the Polar Libraries Colloquy (PLC), an international forum for addressing the collection, preservation, and dissemination of polar information. Over time PLC developed an online directory of polar-related libraries, including their unique subject strengths and current contact and access information. The directory is maintained by the Scott Polar Research Library (SPRI) at the University of Cambridge, United Kingdom. It is now known as the SPRI Polar Directory and is available at http://www.spri.cam.ac.uk/resources/directory/libraries/.

A similar directory for marine libraries includes polar marine collections and is made available through another professional organization, the International Association of Aquatic and Marine Science Libraries and Information Centers (IAMSLIC) at http://www.iamslic.org/publications/inventory-of-marine-and-aquatic-repositories. Searching through these directories allows researchers to identify libraries or institutions with collections that match their research interests.

Library Catalogs

Once students have located polar libraries and catalogs, they need to employ the skills to search those resources effectively. While keyword searching will reveal some polar materials, exploiting subject headings and other indexing will give a more sophisticated level of search. Library catalogs with LC Subject Headings (LCSH) offer entry points into the published and collected literature. Both of the terms *Antarctica* and *Arctic Regions* are subject headings, but they can also be geographic headings, so it will be helpful for students to learn to be aware of this when searching. Table 6.1 shows examples of searches that, when used in the subject field of a database or library catalog employing LCSH, will retrieve relevant materials.

For an example of an information literacy assignment created to teach the use of LCSH in searching, see Polar Information Activity 2 in the Polar Information Literacy Instruction section.

Table 6.1 Examples of searches using Library of Congress Subject Headings terms

Animals—Polar Regions	Paleobotany—Franz Joseph Land
Entomology—Antarctic regions	Reindeer—Finland
Habitat conservation—Antarctica	Tundra plants—Arctic regions
Lichens—South Orkney Islands—Signy Island	White whale hunting—Greenland

Expedition Journals and Accounts

Expedition accounts are one of the most important resources for the Polar Regions. Many of the explorers and adventurers kept diaries or returned home to write accounts of their travels. Many Arctic and Antarctic exploration teams also included a team doctor, botanist, and/or zoologist. These scientists recorded all of the new species they encountered, made drawings or photographs, and often returned with specimens. These are often the earliest accounts of the flora and fauna existing in the Arctic and Antarctic.

Exploration journals, diaries, photographs, audio and video materials, and maps can be found in the general collections and rare book rooms of most polar libraries. The Scott Polar Research Institute in Cambridge, the Byrd Polar Center in Ohio, and Arktikum in Finland, for example, all hold exploration journals; many more libraries with similar holdings are listed in the SPRI Polar Directory. Exploration collections are also being scanned and made available electronically. For example, Google has digitized many exploration works and made them available through Google Books. Numerous other digitization activities are underway, including the OA Biodiversity Heritage Library at http://www.biodiversitylibrary.org/, which serves as the foundational literature component of the Encyclopedia of Life.

For an example of an information literacy assignment that involves searching the Biodiversity Heritage Library to find Antarctic species information, see Polar Information Activity 3.

Reference Tools

Among the resources that students will be able to find through catalogs and in library collections are standard reference sources. The following listing gives representative examples of reference tools focused on the Polar Regions.

Polar Information Activity 4 provides an example of an information literacy assignment that involves selecting and searching a variety of these standard reference sources.

Atlases, Maps, and Polar Mapping Sites

Pan Inuit Trails Atlas

Focusing on eastern and western Canadian Arctic initially, this project draws on historical written records and maps to show Inuit occupancy and mobility.

http://www.paninuittrails.org/index.html?module=module.intro.

Arctic Environmental Atlas, UN Environment Programme/ GRID-Arendal

This interactive atlas of the Arctic presents environmental features such as land cover and land use, topography, wilderness, population density, and protected areas.

http://maps.grida.no/arctic/.

Toolik-Arctic Geobotanical Atlas (TAGA), University of Alaska Fairbanks

TAGA features geobotanical maps and other materials presented at varying scales. Data focus on two research sites in Alaska but also extend to the Circumpolar Arctic. The atlas is part of the Greening of the Arctic Initiative of the IPY.

http://www.arcticatlas.org/.

Annual Arctic Ice Atlas

Since 1990 the Canadian Ice Service has presented an annual atlas documenting Arctic winter sea ice conditions in Canada, all of which are accessible from this site.

https://www.ec.gc.ca/glaces-ice/?lang=En&n=03094EB3-1.

Antarctic Research Atlas
Developed and made available by the US Geological Survey (USGS), this atlas serves as an information framework and research aid to Antarctica and its geospatial data. It is part of the larger cooperative Landsat Image Mosaic of Antarctica (LIMA) project.
http://lima.usgs.gov/antarctic_research_atlas/.

Encyclopedias

Nuttall, M. (2005). *Encyclopedia of the Arctic*, (print). Routledge.

Hund, A. (2014). *Antarctica and the Arctic Circle: A geographic encyclopedia of the Earth's Polar Regions*, (2 print vols. or electronic). ABC-CLIO.

Dictionaries

A number of dictionaries, both thematic and linguistic, are specific to the Polar Regions:

Fortescue, M. D. (2010). *Comparative Eskimo dictionary: With Aleut cognates*. Fairbanks, AK: Alaska Native Language Center, University of Alaska Fairbanks.

Hince, B. (2000). *The Antarctic dictionary: A complete guide to Antarctic English*. UNESCO.

Nordic Cooperation: Nordic Dictionaries
Multiple dictionaries are available through this site, with many freely accessible.
http://www.norden.org/en/fakta-om-norden-1/nordic-dictionaries.

Glosbe English-Northern Sami Dictionary
This freely accessible English-Northern Sami dictionary is an open, collaborative effort.
https://glosbe.com/en/se/.

Inuktitut Living Dictionary
An open, collaborative dictionary from Nunavut, Canada, this project encourages Inuktitut speakers in all communities to contribute.
http://www.livingdictionary.com/.

Polar Journals

Once students have found materials readily available in library catalogs, they need to learn how to locate articles in polar journals. It is helpful to be familiar with the titles of some of the journals specific to the Polar Regions as listed below. Many of the polar research titles will contain the words Antarctic and/or Arctic. Polar will also often appear in a title, as will other identifying words referring to the regions' geography or Indigenous peoples and cultures.

Antarctic; Antarctic Journal; Antarctic Journal of the United States; Antarctic Research Series; Antarctic Science; Arctic, Antarctic & Alpine Research; Arctic; Arctic Anthropology; Arctic Journal; Berichte zur Polar- und Meeresforschung; Études/Inuit/Studies; Hemispheric and Polar Studies Journal; International Journal of Circumpolar Health; The Northern Review; Ocean and Polar Research; Polar Biology; Polar Bioscience; Polar Geography; Polar Record; Polar Science; and *Polish Polar Research.*

Polar Databases

Many polar journals are indexed in standard resources, but some indexes focus entirely on polar publications or contain significant amounts of polar information. These include the following:

Arctic and Antarctic Regions
Priced publication, check your library for availability.
Antarctic and Cold Regions Bibliographies
http://glossary.agiweb.org/dbtw-wpd/cold/coldA.htm.
Arctic Science and Technology Information System, ASTIS
http://www.aina.ucalgary.ca/astis/.
e-Library.ru
http://elibrary.ru/defaultx.asp.
High North Research Documents
http://highnorth.uit.no/.
Hubert Wenger Eskimo Database
http://wenger.library.uaf.edu/.
iPortal: Indigenous Studies Portal Research Tool
From the University of Saskatchewan, this full-text database focuses on First Nations and Aboriginal peoples of Canada. The site includes iPortal News, a brief current news service.
http://iportal.usask.ca/.
International Polar Year Publications Database
IPYs are scheduled, collaborative, and international efforts during which intense research is conducted on the Polar Regions. The first occurred in 1882–83, and the second IPY took place in 1932–33. The International Geophysical Year, also referred to as the third IPY, occurred in 1957–58. The fourth and most recent IPY transpired over two full annual cycles from 2007 to 2009.
Priced publication from NISC; check your library for availability.

Lapponica
This is the joint database for the libraries of Lapland.
http://intro.rovaniemi.fi:8002/Intro?formid=lapf2&ulang=eng.
Russian Web of Science
Priced publication; check your library for availability.
SPRILIB Russian North
The SPRI (Scott Polar Research Institute) Library "supports research into the Polar Regions and ice and snow wherever found." It has an extensive collection of publications concerning northern Russia and Siberia dating from 1671 through to 2003 and is periodically updated. http://www.spri.cam.ac.uk/library/catalogue/russian/.

Government Publications

All of the countries with polar interests have established related departments or agencies, many with active publication programs, and often libraries, archives, and data and specimen repositories. For countries with Antarctic interests, the agencies are obvious. These include, for example, the British Antarctic Survey available at https://www.bas.ac.uk/, Japan's National Polar Research Institute available at http://www.nipr.ac.jp/english/antarctic/, and the New Zealand Antarctic Research Institute available at http://nzari.aq/. Examining the websites of these and other agencies reveals rich information sources.

Arctic countries all have direct legislative responsibility for their Arctic lands and waters, so within their normal government publication processes there are Arctic-related publications. Some polar government resources appear in government publications indexes, such as the Catalog of U.S. Government Publications available at https://catalog.gpo.gov/F and the Government of Canada Publications Catalogue available at http://publications.gc.ca/site/eng/ourCatalogue.html. However, government databases rarely index all publications of government agencies, so often researchers must take the next step to find websites for individual departments within the agencies and search each one for their publications.

This is the case for the Norwegian Environment Agency available at http://www.miljodirektoratet.no/en/Publications1/. Another example is the Alaska Department of Fish & Game, which publishes reports on its website available at http://www.adfg.alaska.gov/index.cfm?adfg=librarypublications. wildlifepublications. It also makes its reports and other materials available through the Alaska Resources Library and Information Services (ARLIS) at http://www.arlis.org/. ARLIS is the depository library for many federal, state, university, and private entities involved in research related to Alaska and the Arctic. Its collections are strong in information concerning oil spills and remediation in cold-water environments (thanks to the 1989 *Exxon Valdez* oil spill that occurred in Alaska's Prince William Sound), as well as in other topics including fish and wildlife, climate, subsistence, ecology, baseline data, land use, oil and gas, and geology.

Documents From Hearings, Commissions, Assessments, etc.

In the Arctic especially, some of the richest information about life science subjects is contained in Indigenous land claim negotiations and hearings, as well as in environmental impact statements and assessments of industrial developments, such as pipelines, tourist service projects, and mines. Frequently these involve depositions related to the environment, which usually include information about biological subjects, but are often difficult to identify. Examples include the following:

MacKenzie Valley Pipeline Inquiry Report
https://docs.neb-one.gc.ca/ll-eng/llisapi.dll?_gc_lang=en&func=ll&objId=238336&objAction=browse&redirect=3.

Alaska North Slope Natural Gas
Hearing before the Committee on Energy and Natural Resources, United States Senate, One Hundred Sixth Congress, second session, to consider the transportation of Alaska North Slope natural gas to market and to investigate the cost, environmental impacts, and energy security implications to Alaska and the rest of the nation for alternative routes and projects, September 14, 2000.
https://catalog.hathitrust.org/Record/010244309.

Nunavik Marine Region Wildlife Board
Southern Hudson's Bay Polar Bear Management [Public Hearings].
http://nmrwb.ca/index.php/en/resources/public-hearings/southern-hudson-bay-polar-bear-management.

International Treaties and Agreements

Parts of the Polar Regions are governed by a variety of treaties and agreements that often address land use, and protection and conservation of biological resources. The Arctic Governance Project has assembled a Compendium of existing and proposed governance arrangements in the Arctic available at http://arcticgovernance.custompublish.com/compendium.137742.en.html. It includes sections for marine, terrestrial, and atmospheric agreements, as well as Indigenous approaches to governance. Important examples include the following:

Antarctic Treaty Database
http://www.ats.aq/devAS/info_measures_list.aspx?lang=e.

Polar Code MECP 68-21 Annex 10
http://www.imla.co/sites/default/files/mepc_68-21_-_report_of_the_marine_environment_protection_committee_on_its_sixty-eighth_session_secretariat.pdf.

Spitsbergen Treaty
https://en.wikisource.org/wiki/Spitsbergen_Treaty.

United Nations Convention on the Law of the Sea of December 10, 1982
http://www.un.org/Depts/los/convention_agreements/texts/unclos/closindx.htm.

Gray Literature

Much of polar literature is considered "gray literature," i.e., literature that is not published through the usual scholarly channels, is often ephemeral, and may or may not be collected by libraries. Gray (or grey) literature is sometimes called "fugitive literature" because it is so frequently hard to find. Some scholars include government publications, particularly the ephemeral ones, in the definition of gray literature (Luzi, 2000; MacDonald, Ross, Soomai, & Wells, 2015).

Teaching students to search for gray literature can be challenging. All students need to be trained to think reflectively about what entities are likely to conduct research and create information regarding the Arctic and Antarctic. This will help them know, at a broad scale, where to begin looking. In more recent years, students have also needed to consider what parties would pay to have polar collections digitized.

Many nongovernment agencies (international, regional, local, academic, private) web-publish or point to biological information as well. In many of these cases, students will need to identify the organization and search its webpages, looking for headings called "publications," "documents," "archives," and/or "reports." Arctic Portal maintains an extensive list of these organizations on its acronyms page at http://arcticportal.org/acronyms. Examples of important polar organizations that make publications available on their websites include the following:

Arctic Council
http://www.arctic-council.org/index.php/en/documents.

Antarctic and Southern Oceans Coalition
http://www.asoc.org/news-and-publications/publications.

Barents-Euro Arctic Council
http://www.beac.st/en/Documents.
Several organizations that index or harvest some of these publications include the following:
Arctic Portal Library
Its purpose is to maintain a comprehensive collection of Arctic-relevant scientific and educational material, especially material produced or published within the scope of IPY 2007–08, the Arctic Council, and the University of the Arctic. It also makes available material produced in various Arctic research programs and projects.
http://library.arcticportal.org/.
High North Research Documents
http://highnorth.uit.no/.
University of Alberta Circumpolar Collection (Web Archive)
https://archive-it.org/collections/2475.

Archival Materials, Including Photos and Videos

Many organizations with polar libraries also have polar archives. A few examples are as follow:
Alaska and Polar Regions Collections & Archives
Elmer E. Rasmuson Library, University of Alaska Fairbanks.
http://library.uaf.edu/apr-collections.
Alaska's Digital Archives
http://vilda.alaska.edu/.
Arctic Institute of North America Photo Collection
http://contentdm.ucalgary.ca/cdm/search/collection/aina.
Barentsinfo.org—Photos
http://www.barentsinfo.org/Photos.
Byrd Polar and Climate Research Center Archival Program
The Ohio State University
https://library.osu.edu/find/collections/byrd-polar-archives/.
Circumpolar Digital Image Collection. ERA: Education and Research Archive
University of Alberta
https://era.library.ualberta.ca/collections/44558t91d.
Freeze Frame: Historic Polar Images, 1845–1982, from Scott Polar Research Institute
http://www.freezeframe.ac.uk/home/home.

Polar Data Repositories

Academic libraries have become involved in data management spurred in part by the need to manage data collected during the most recent IPY. Several polar data sources are available, including the following:

Polar Data Catalogue
Canadian Cryospheric Information Centre
Data and metadata sources included here are for both the Arctic and the Antarctic.
https://www.polardata.ca/pdcsearch/.

International Polar Year Historical Data and Literature
National Snow and Ice Data Center
http://nsidc.org/data/docs/noaa/g02201_dahli/.

Meerisportal.de Data Portal
http://data.seaiceportal.de/gallery/index_new.php?lang=en_US.

UK Polar Data Centre
https://www.bas.ac.uk/data/uk-pdc/.

Traditional Knowledge

For the Arctic, the traditional knowledge (TK) of Indigenous peoples is an important source of information; in fact, it is often the best or only information available on a subject. Some TK has been captured in the form of videos and field studies and even published as books or series. The material can be open to the public or access to it can be restricted to community members. Sometimes access needs to be negotiated with the Indigenous owners of the information. Often the information has not been documented, requiring researchers to ask elders and other community members directly for stories or testimonials.

As a part of information literacy training, it is important for students to learn that there are often protocols to be followed when approaching Indigenous communities for TK. Some organizations have worked with Indigenous communities to develop protocols for interactions that include Arctic Indigenous people. For example, Nunavut Arctic College describes how researchers should approach elders in its Guidelines for Working with Inuit Elders, available at http://www.arcticcollege.ca/publications/books/elder%20guide%20english.pdf.

For some Indigenous peoples, their knowledge may be sacred or even "owned" by an elder, and they will not disclose it inappropriately. In other situations, cultural sensitivities may cause people to not share information.

For example, the location of sacred sites, images of the dead, or the preparation method for a traditional medicine may be things people do not wish to reveal. Even if community members were willing to disclose information, they may choose not to do so if they think that the information will be insensitively used, abused, or otherwise not respected. As a part of the process of accessing TK as a source of information, it is important for students and other researchers to take the time to learn about how the information is valued and how to properly offer respect within particular communities. In the absence of a published protocol, it is usually appropriate to simply ask a community contact how one should offer respect.

DISCIPLINE INFORMATION LITERACY INSTRUCTION

When teaching information literacy for any discipline, it makes sense to work from known to unknown, simple to complex, easy to difficult, and concrete to abstract. When teaching polar information literacy, begin with sources that users are likely to know: library catalogs, Google or Google Scholar, and tertiary resources such as encyclopedias, dictionaries, and handbooks. Build from this base of knowledge to more unusual, difficult, and obscure resources.

In the field of information literacy, an inquiry-based learning approach followed by student engagement and student-centered learning is an important theme (Ross & Furno, 2011; Stegman, 2014). It is no less important in polar information literacy. Librarians are always inventing new and interesting ways of engaging learners. The Alfred Wegner Institute created an appealing YouTube video, "How does the polar bear Bert find a new home?" available at https://www.youtube.com/embed/dztwrdvZj_c, that encourages people to use their sea ice information portal, Meerisportal.de at http://www.meereisportal.de/. The Scott Polar Research Institute created a game in which participants are given a collection of cards featuring pictures of members of historical Antarctic expeditions on the front and facts about them on the back; players must select a good expeditionary team (Amos & Chapman, 2014).

For information literacy to be effective, it needs to answer the "what's in it for me" question from the learner. We know that students are more likely to integrate the knowledge if they have an immediate application for the skills that they learn (Banas, 2009). Ideally, polar information literacy will be embedded into courses in which students have research-related assignments or into graduate studies programs in which research is the learner's purpose. In the following section are examples of practical polar assignments, which could be incorporated into coursework. The information activities have been referred to in appropriate contexts throughout the text above. The example questions only appear below.

Examples of Polar Information Literacy Assignments
Polar Information Activity 1
Find polar projection maps for the North and South Poles. Identify the countries around the Arctic Ocean and the primary countries involved in the Antarctic Treaty and Antarctic research.
- Example for the Arctic: http://ftp.geogratis.gc.ca/pub/nrcan_rncan/raster/atlas/eng/reference/circumpolar/MCR0001_circumpolar_2008.jpg.
- Example for Antarctica: https://www.comnap.aq/Publications/Comnap%20Publications/COMNAP_Map_Edition4_A0_2009-03-26.pdfhttps://www.comnap.aq/Publications/Comnap Publications/COMNAP_Map_Edition4_A0_2009-03-26.pdf.

Polar Information Activity 2
Select a marine species (e.g., salmon, whales, seals, penguins, puffins) and search for it in one of the following library catalogs, using the species as a subject heading along with Antarctica, Arctic Regions, or a polar country or state name as a subheading. Give examples, i.e., Whales—Iceland, Walrus—Alaska.
- Union Catalogue of Icelandic Libraries (Leitir.is)
  http://leitir.is/primo_library/libweb/action/search.do.
- SPRILIB Antarctica
  http://www.spri.cam.ac.uk/library/catalogue/antarctica/
- University of Lapland Library Services
  https://luc.finna.fi/ulapland/Search/Advanced
- University of Alaska Fairbanks Library Catalog
  http://catalog.library.uaf.edu/uhtbin/cgisirsi/?ps=X8FatWgGMF/UAFRAS/96570002/38/1/X/BLASTOFF
- The Roger G. Barry Archives and Resource Center (ARC) at NSIDC
  https://nsidc.org/fmi/iwp/cgi?-db=NSIDC_Library&-loadframes

Polar Information Activity 3
Using five sources in the Biodiversity Heritage Library at http://www.bhlscielo.org/, describe five different plant species found in the Antarctic. Answer the following questions:
- What is the genus name?
- Where were the species found in the Antarctic?
- Who found them?
- Which expedition(s)?
- What are the distinguishing features, if any?

Polar Information Activity 4
Select one of the following species that occurs in the Arctic or Antarctic: caribou/reindeer, beluga/narwhal, polar bear, harp seal, whooping crane, tundra swan, ptarmigan, any species of penguin, Atlantic cod, etc. Find authoritative sources/complete the activities for the following:
- Description of natural history of the species.
- Current status.
- Impact of global warming/climate change (at least x number of books, journal articles, theses or dissertations, etc.).
- Associations working specifically with the species that have published information about that species.
- Publicity, newspaper articles about the species.
- Information quoting local people/elders on the state of the species (if in a populated area).
- OA image showing some behavioral aspect of the species (not just a photograph of the species).
- Video that shows some behavioral aspect of the species in the wild.
- Post on a blog.
- Collaborate with another student to write a two-page fictional account involving the species that each of you researched; the account must incorporate accurate details about these animals.
- Cite each of your sources correctly and consistently in one style.

Example Questions for Information Literacy Assignments
- Using the High North Research harvester, find five different kinds of materials (e.g., conference presentation, map, dissertation or thesis, research paper) related to Sami reindeer herding and climate change.
- Using the Scott Polar Research Institute's photograph collection, the Byrd Polar Research Center's collection, or the British Antarctic Survey's library, find information about the use of animals (dogs, cows, and horses) in the exploration of Antarctica.
- Smoking of fish is a food preservation practice employed around the Circumpolar North. Use library catalogs, journal articles, dissertations and theses, and personal accounts to find information about fish-smoking techniques.
- On Friday, September 16, 2016, the first commercial cruise ship stopped at Canada's Cambridge Bay on its way through the Northwest Passage. Find studies, including government publications, relating to the potential environmental impact of regular cruise ship movements through the Northwest Passage.

- Use the library of the Australian Antarctic Survey to find information about seabird surveys on the Antarctic Coast.
- Use the library of the Alfred Wegner Institute to find reports about Greenland's sea ice changes and impact on polar bears.
- Find data related to permafrost melt collected at the Abisko Research Station in northern Sweden.
- During the IPYs, much data were collected. Locate IPY data collected around Svalbard, Norway, and determine how to access it.
- Blueberries grow all around the Circumpolar North. In addition to their nutritional value, they are also considered to be a medicinal food in many Indigenous cultures. Find information about the traditional uses of blueberries.
- Seal hunting provides a traditional and healthy source of food for many Indigenous northern peoples; however, it is also controversial. Find Canadian newspaper sources that describe the controversy.
- Consult Indigenous dictionaries to find the word for "seal" in three different polar languages.
- Find a first-hand account of life in a traditional Inuit fish camp.
- Find an Indigenous-authored children's book about Arctic wildlife.
- Find information about the migration of swans from Japan to northern Russia.
- The Ainu are northern Indigenous people of Japan. Find information on their traditional use of whales.
- Your thesis will be on the viability of outdoor gardening in Arctic regions, with attention to the kinds of garden plants that are particularly successful. Find materials, including government publications, that address this subject.

EMERGING TRENDS AND STAYING UP-TO-DATE
Digitization of Polar Collections

Many organizations are in the process of digitizing their polar collections, focusing on unique items or items that directly support their mandates. For example, the Inuvialuit Cultural Resource Centre is currently in the process of digitizing many of its records. These projects are sometimes piecemeal and funded by short-term grants. In other cases, they are government-sponsored initiatives aimed at digitizing collections of thousands of items. Over time, more of the polar record is becoming digitally accessible. Information literacy programs will need to educate students not only about how to identify and access these resources but also about how to not become overwhelmed by the volume of available material.

Open Access

The movement toward making research more accessible has been growing, with more government and funding agencies mandating that researchers' published findings are made OA. Agencies requiring OA include the Australian Research Council (2013), Research Councils UK (2014), Canada's national Tri-Council Agencies (2015), and many more. There are several kinds of OA resources, such as peer-reviewed journals or articles that are either free to readers from the point of publication by major publishers, known as "gold OA," or those that are self-archived using institutional or subject repositories, called "green OA." Gold OA requires author processing fees, making it costly for authors; however, based on licensing and copyright restrictions, this is sometimes the only solution for making OA work. The situation is complicated by many predatory publishers taking an author's money to publish but supplying no quality control or peer review (Zhao, 2014). This makes it especially important for information literacy programs to educate students about how to distinguish high-quality polar works from those that are not.

Scholarly Open Access, available at https://scholarlyoa.com/, is a website that lists journals that students should examine closely for quality; 10 of the titles there begin with the word Antarctic. The Directory of Open Access Journals, available at https://doaj.org/, offers access to peer-reviewed OA publications, including titles on Arctic, Antarctic, and polar research.

Increased Availability of Polar Materials in Many Languages

While English is the most common language of scientific communication, there has always been non-English publication about the Polar Regions. Fifty-three countries, with various languages, have signed the Antarctic Treaty to date. In the Canadian Arctic, at least 15 Indigenous languages and many more dialects are spoken. With the lessening of the Arctic Ocean ice, many more nations are seeking toeholds in the Arctic through participation in Arctic research, as well as business and trade activities. Over time as the Internet and its network services have revolutionized and democratized the publication and distribution of information, non-English polar materials have become much more readily available. For example, Inhabit Media of Iqaluit, Canada, routinely publishes children's works, many of them about plants and animals, simultaneously in Inuktitut and English. Automatic translation software is becoming ubiquitous, and researchers in polar biology will expect to be able to use these programs. Information literacy programs will need to include not only information about translation options but also critical evaluation skills for machine-translated materials.

Current Awareness Services

Many research stations and polar organizations have their own newsletters, blogs, and social media feeds. One example is the International Polar Foundation's RSS Feed for the Princess Elisabeth Antarctic Research Station available at http://www.antarcticstation.org/rss. Several services collate news about the Arctic and the Antarctic. Some examples are as follows:

- Alaska Dispatch News at https://www.adn.com/.
- The Arctic (Russian Geographical Society) at http://arctic.ru/.
- Arctic News at http://arctic-news.blogspot.ca/.
- Arctic Update at https://www.arctic.gov/arctic_update/index.html.
- Eye on the Arctic at http://www.rcinet.ca/eye-on-the-arctic/.
- High North News at http://www.highnorthnews.com/.
- Iceland Monitor at http://icelandmonitor.mbl.is/.
- University of the Arctic News at http://www.uarctic.org/news/.

CONCLUSION

Search engines such as Google can almost instantaneously retrieve a beguiling volume of materials based on cursory searches. To return to the metaphor of the ice-covered lake, Google retrieval is the thin, surface veneer of the easily identifiable information. It is our job as instruction librarians to lead students, researchers, and the public to the troves of information that can be found in the depths of the Deep Web. Of course, we must teach the standard sources in life sciences that will retrieve scholarly publications related to the Polar Regions. However, for more extensive and difficult searches, we must teach a process that encourages students and other researchers to think through from whence the information could have originated and to begin their searches there.

REFERENCES

ALA/ACRL/STS Task Force on Information Literacy for Science and Technology. (2006). *Information literacy standards for science and engineering technology*. Retrieved from http://www.ala.org/acrl/standards/infolitscitech.

Amos, R., & Chapman, N. (July 2014). *Using polar library and archive resources for public outreach*. Unpublished paper presented at the meeting of the 25th Polar Libraries Colloquy, held 29 June–3 July 2014, Cambridge, United Kingdom.

Association of College and Research Libraries [ACRL]. (2000). *Information literacy competency standards for higher education*. Retrieved from http://www.ala.org/acrl/standards/informationliteracycompetency.

Association of College and Research Libraries [ACRL] Board. (2016). *Framework for information literacy for higher education*. Retrieved from http://www.ala.org/acrl/standards/ilframework.

Australian Research Council. (2013). *ARC open access policy*. Retrieved from http://www.arc.gov.au/arc-open-access-policy.

Banas, J. R. (2009). Borrowing from health communications to motivate students to learn information literacy skills. *Community & Junior College Libraries, 15*(2). http://dx.doi.org/10.1080/02763910902832214. 18; 82.

Campbell, S. (2008). Chapter I – defining information literacy in the 21st century. In S. Koopman, & J. Lau (Eds.), *IFLA Publications 131. Information literacy: International perspectives* (pp. 17–26). Munich: K. G. Saur.

Fabbi, J. L. (2012). *Fortifying the pipeline: An exploratory study of high school factors impacting the information literacy of first-year college students*. Doctoral dissertation UNLV [University of Nevada, Las Vegas]. Theses, Dissertations, Professional Papers, and Capstones, 1516. Available from http://digitalscholarship.unlv.edu/thesesdissertations/1516.

Garcia, L. (2014). Applying the framework for information literacy to the developmental education classroom. *Community & Junior College Libraries, 20*(1–2), 39–47. http://dx.doi.org/10.1080/02763915.2014.1013399.

Luzi, D. (2000). Trends and evolution in the development of grey literature: A review. *International Journal on Grey Literature, 1*(3), 105–116. http://dx.doi.org/10.1108/14666180010345537.

MacDonald, B. H., Ross, J. D., Soomai, S. S., & Wells, P. G. (2015). How information in grey literature informs policy and decision-making: A perspective on the need to understand the processes. *The Grey Journal, 11*(1), 7–16. Retrieved from http://search.proquest.com/docview/1703151412?accountid=14474.

Mackey, T. P., & Jacobson, T. E. (2014). *Metaliteracy: Reinventing information literacy to empower learners*. Chicago: Neal-Schuman.

Research Councils UK. (2014). *RCUK policy on open access*. Retrieved from http://www.rcuk.ac.uk/research/openaccess/policy/.

Ross, A., & Furno, C. (2011). Active learning in the library instruction environment: An exploratory study. *Portal: Libraries and the Academy, 11*(4). 18; 970. Retrieved from http://search.proquest.com.login.ezproxy.library.ualberta.ca/docview/1013807451?accountid=14474.

Scaramozzino, J. M. (2008). An undergraduate science information literacy tutorial in a web 2.0 world. *Issues in Science and Technology Librarianship* (55). http://dx.doi.org/10.5062/F4VM4960.

Secretariat of the Antarctic Treaty. (2017). *ATS: The Antarctic Treaty*. Retrieved from http://www.ats.aq/e/ats.htm.

Stegman, B. (2014). Inquiry, new literacies, and the common core. *Kappa Delta Pi Record, 50*(1). http://dx.doi.org/10.1080/00228958.2014.871688. 6; 36.

Tri-Council Agencies, Canada. (2015). *Tri-agency open access policy on publications*. Retrieved from http://www.science.gc.ca/eic/site/063.nsf/eng/h_F6765465.html?OpenDocument.

UDC Consortium. (2017). *About universal decimal classification (UDC)*. Retrieved from http://www.udcc.org/index.php/site/page?view=about.

Zhao, L. (2014). Riding the wave of Open Access: providing library research support for scholarly publishing literacy. *Australian Academic & Research Libraries, 45*(1), 3–18. http://dx.doi.org/10.1080/00048623.2014.882873.

CHAPTER 7

Zoology and Animal Sciences Information Literacy

Daria O. Carle
University of Alaska Anchorage, Anchorage, AK, United States

INTRODUCTION

Whenever anyone has asked a question or wondered about an animal, what it eats or where it lives, zoological information has provided the answer. From children in elementary and middle schools who read books about their favorite animal, to high school and college students who write research papers on more complicated topics related to animals—they are all studying zoology. Graduate students and faculty in biology departments, academic researchers, scientists at research institutes, zookeepers and zoologists, veterinarians, ichthyologists, ornithologists, mammologists, animal behaviorists, geneticists, physiologists, and ecologists all need to know about zoology.

The *Oxford English Dictionary* defines zoology as the "scientific study of the behaviour, structure, physiology, classification, and distribution of animals" (Zoology, 2017). For the purposes of this chapter, the term zoology refers to the biology of animals in the broadest sense, from the entire ecosystem to the whole organism at the macrolevel. Subdisciplines within zoology include anatomy, physiology, genetics, ecology, natural history, conservation, animal behavior, and environmental science, to name just a few.

The history of zoology dates back to the earliest times. "People have been studying animals since the dawn of time although the intent and the degree of scientific validity of their conclusions have varied over the ages" (Schmidt, 2003, p. 2). Over the centuries, many discoveries and developments have furthered the study of zoology, but one significant milestone along the way was the development of the foundational system for how the animal kingdom is organized.

TAXONOMY/SCIENTIFIC NOMENCLATURE

In 1735, Swedish scientist Carl Linnaeus created the binomial nomenclature system that provided the basis for naming the plants and animals of the natural world. Linnaeus' system formalized all living things based on taxonomy, i.e., the science of naming, describing, and classifying groups of organisms on the basis of shared characteristics and giving names to those groups (Convention on Biological Diversity, n.d.). Linnaeus organized names in a hierarchical structure, as in kingdom, phylum, class, order, family, genus, and species. Over the years, a number of mnemonic phrases have been invented to help remember this taxonomic scheme, KPCOFGS, such as "kids prefer cheese over fancy green salad" and "King Philip came over for great soup." One favorite, in keeping with the theme of this book, is "keeping precious creatures organized for grumpy scientists."

In this way, an individual species is represented by its scientific name within a hierarchical structure. For example, the taxonomic classification of a tiger is:

Kingdom	Animalia
Phylum	Chordata
Class	Mammalia
Order	Carnivora
Family	Felidae
Genus	*Panthera*
Species	*tigris*

The genus, always capitalized, and species names both appear in italics. Some species are further represented by a subspecies, also italicized, that is frequently used to distinguish one geographical distribution of occurrence from another. Finally, a surname, and often the date (both in parentheses), of the individual or group who first identified or named the organism is listed. In this way, each individual species has a unique and distinct scientific name. To continue the example, a Siberian tiger's full binomial name is *Panthera tigris altaica* (Temminck, 1844) and is exclusive to that species.

On the other hand, animals often have many common names that vary widely based on culture, language, and locale, making a search for information on that animal somewhat problematic if only a colloquial term is used. For this reason, the recommended practice is to use both the scientific (e.g., genus and species) and common name(s) of an organism when searching for information on a particular species, particularly in science-focused subject resources and databases.

DISCIPLINE RESOURCES

The field of zoology is broad and diverse, and so too are its information resources. From reference books to handbooks and field guides, to journals, databases, and institutional repositories, there is a plethora of information containing information related to zoology. What follows is a selected list of resources, with descriptions adapted from the websites, publishers, and database providers. For a more detailed and comprehensive coverage of the zoology discipline, consult D. Schmidt's *Guide to Reference and Information Sources in the Zoological Sciences* (2003) and related website at http://www.library.illinois.edu/bix/zoology/.

Taxonomic Information

Integrated Taxonomic Information System (ITIS)
A partnership of North American agencies, organizations, and taxonomic specialists, ITIS is the authoritative taxonomic information on the plants and animals of the world and provides the basis for the Encyclopedia of Life. https://www.itis.gov/.

International Union for Conservation of Nature (IUCN) and Natural Resources—Red List of Threatened Species
This resource describes taxonomic, conservation status, and distribution information on plants, fungi, and animals that have been globally evaluated using the IUCN Red List Categories and Criteria. It ranks the relative risk of extinction by highlighting those plants and animals at a higher risk of global extinction and also lists species that are classified as extinct or extinct in the wild.
http://www.iucnredlist.org/.

Encyclopedias and Reference Books

Far too numerous to name, encyclopedias and reference books on zoology, animals, and related topics abound, in both print and online form. They consist of general, multivolume works such as *Grzimek's Animal Life Encyclopedia* that covers the entire animal kingdom, while others are very specialized and feature a specific species or group of species, e.g., *New Encyclopedia of Snakes*. Field guides are reference books that are used as a tool to identify a species by particular characteristics, and they often describe a particular group of species in a specific geographic location, e.g., *The Birds of Panama: A Field Guide*. Examples of online encyclopedias in zoology include the following:

Encyclopedia of Life (EOL)
The EOL, a project begun in 2007 to "provide a webpage for every species," has developed into a community of collaborators worldwide who contribute information to the resource. Sustained by grant funding and contributions from supporting institutions, EOL contains information and pictures of all species known to science.
http://eol.org/.

United Nations Educational, Scientific, and Cultural Organization (UNESCO)—Encyclopedia of Life Support Systems (EOLSS)
Developed under the auspices of UNESCO, EOLSS is regarded as the world's largest comprehensive publication of life on planet Earth. This high-quality peer-reviewed compendium of 21 encyclopedias is thematically organized, and each covers archival content in many traditional disciplines and interdisciplinary subjects. It also includes up-to-date coverage of various aspects of sustainable development that are relevant to the current state of the world.
http://www.eolss.net/.

Animal Diversity Web
An educational resource from the University of Michigan, Animal Diversity Web is an online database and encyclopedia of animal natural history for exploring biodiversity. Built through contributions from students, photographers, and many others, it is designed for use in inquiry-based education and supports the learning experiences of students in zoology, evolution, behavior, and ecology courses.
http://animaldiversity.org/.

Directory of Open Access Books (DOAB)
Developed as a means to increase availability of Open Access, peer-reviewed books, DOAB records can also be integrated into library online catalogs, thereby helping scholars and students to discover the books included in this resource. The option to browse by subject is included.
http://www.doabooks.org/doab?func=subject&uiLanguage=en.

Databases

A number of specialized abstracting and indexing databases that cover the subject of zoology exist, but the following are considered core. Google Scholar and PubMed are freely available on the Web, and the rest are subscription-based resources available in many academic libraries. Descriptions are adapted from the databases' websites.

Google Scholar
An easy way to search for scholarly literature from articles, theses, books, and abstracts from academic publishers, professional societies, online repositories, and other websites, Google Scholar searches for relevant work across the world of scholarly research. It also identifies related works and publications, and the number of times a publication has been cited. Searching by subject is not an option, so using the most

appropriate keywords is important to locate needed information. Using the Advanced Search option allows for entering a more sophisticated search string.
https://scholar.google.com/.

PubMed
A free resource developed by the US National Library of Medicine (NLM) at the National Institutes of Health, PubMed includes the largest collection of biomedical literature in the world. While the primary focus is medicine and health, PubMed also includes portions of the life sciences, behavioral sciences, chemical sciences, and bioengineering. It draws content from MEDLINE (NLM's premier bibliographic database), life sciences journals, and online books, and it provides links to full-text content from PubMed Central and publisher websites. PubMed also includes links to other molecular biology resources from the National Center for Biotechnology Information and additional relevant websites.
https://www.ncbi.nlm.nih.gov/pubmed.

BIOSIS Previews
The standard bearer for information in the life sciences discipline is BIOSIS Previews, an index to the published literature from 1926 forward that includes citations and abstracts to journal articles in all disciplines of biology. It also contains content from the print *Biological Abstracts*, which was formed by the union of *Abstracts of Bacteriology (1917–25)* and *Botanical Abstracts (1919–26)*. Content also includes meetings and conferences, literature reviews, US patents, books, software, and other media from Biological Abstracts/RRM (Reports, Reviews, Meetings) (Sheehy, 1986).

One unique feature of searching BIOSIS Previews is Concept Codes, a five-digit code that represents broad subject areas and taxonomic classification in the life sciences. The codes cover everything from methods, microscopy, and genetics to body systems, physiology, and geological periods. Related Concept Codes are grouped under broader concept headings, e.g., Chordata (vertebrates) includes Concept Codes 62502–62520 covering general/systematic information, Prochordata, Pisces (fish), Amphibia, Reptilia, Aves (birds), through Mammalia. Concept Codes can be searched with a wild card character; for example, 625★ will retrieve all of Chordata.
http://thomsonreuters.com/en/products-services/scholarly-scientific-research/scholarly-search-and-discovery/biosis-previews.html.

Zoological Record
Considered to be the leading taxonomic reference source in animal biology, this is the world's oldest ongoing database. Begun in 1864 by the Zoological Society of London as a print publication, *The Record of Zoological Literature*, it was renamed Zoological Record in 1870 and is now online. Primarily indexing academic publications, Zoological Record also includes books, reports, and meetings. The range of coverage is broad and includes everything from biodiversity of the environment to taxonomy and veterinary sciences.

Zoological Record also serves as the world's unofficial register of scientific names in zoology since it indexes approximately 90% of the world's zoological literature. Some unique features of the database include the ability to determine the first appearance of an animal in published literature, track changes in the classification and relationships of organisms, and check for descriptions of new species.

Published by Thomson Reuters.

Natural Science Collection
This collection of online databases, originally developed as specialized print indexes by Cambridge Scientific (now ProQuest), covers topics across the entire spectrum of the natural sciences. It uses a thesaurus and contains content type ranging from scholarly journals and government publications to conference proceedings, newspapers, and audio/video sources. It consists of three main database groups, Agricultural and Environmental Science; Biological Science; and Earth, Atmospheric, & Aquatic Science; each includes more specific subject database subgroups. http://www.proquest.com/products-services/natural_science.html.

Web of Science
While not strictly a science resource, since humanities and the social sciences are also included, this interdisciplinary database indexes most of the core titles in scientific and technology journal publishing. In addition to analysis tools and information on emerging trends, it also includes data, books, proceedings, and patents and features the ability to search cited references.
http://thomsonreuters.com/en/products-services/scholarly-scientific-research/scholarly-search-and-discovery/web-of-science.html.

Wildlife & Ecology Studies Worldwide
Offering a global perspective on wildlife and ecology, this resource is considered the world's most comprehensive index to literature on wild mammals, birds, reptiles, and amphibians. With a broad range of coverage

on a variety of topic areas, it consists of records from Wildlife Review Abstracts, formerly *Wildlife Review*, a print publication produced by the US National Biological Service until 1996.
https://www.ebscohost.com/academic/wildlife-ecology-studies-worldwide.

Environment Index
Formerly known as Environmental Knowledgebase OnLine, this database offers extensive coverage in the areas of ecosystem ecology and related areas of study from thousands of journals dating back more than 100 years. Special features include searchable cited references, an in-depth environmental thesaurus, and thousands of author profiles.
https://www.ebscohost.com/academic/environment-index.

Journals and Journal Abbreviations

Academic journals related to zoology and the animal sciences number in the thousands and are published all over the world. Since the first appearance of the *Philosophical Transactions of the Royal Society* (London) in 1665, scientific journals have proliferated and cover every aspect of the zoology discipline. Previously, published lists of zoological journals were found in books or database publications, but most are now online. The University of Illinois at Urbana–Champaign's Biotechnology Information Center maintains a detailed list of sources that identify journal titles, along with their abbreviations, at http://www.library.illinois.edu/biotech/j-abbrev.html.

Open Access Journals

As discussed above, there are thousands of journals that contain articles relevant to zoology. Although the majority of academic journals have restricted access because they are behind the firewall of library and institutional subscriptions, there is a growing movement toward Open Access. The following are a few examples.

PLoS (Public Library of Science)
PLoS was founded in 2001 as a nonprofit Open Access publisher, also acting as an advocacy organization to transform the progress of research communication in science and medicine. It has since expanded to include additional peer-reviewed journals across other areas of science and medicine, including PLoS Biology, PLoS Genetics, and PLoS Medicine.
https://www.plos.org/.

Directory of Open Access Journals (DOAJ)
DOAJ began in Sweden in 2003. Like the DOAB, it covers a broad range of topics in all areas of science, technology, and medicine, as well as the social sciences and humanities. It functions as a community-curated index and directory for high-quality, Open Access peer-reviewed journals, with an option to limit by subject.
https://doaj.org/.

PubMed Central
Launched in 2000, this database is a free full-text archive of biomedical and life sciences journal literature from the National Library of Medicine, part of the US National Institutes of Health.
https://www.ncbi.nlm.nih.gov/pmc/.

Many of the scientific publishers also offer Open Access to selected journals; some immediately and others after a specified embargo period. Here are a few examples.

Elsevier Open Archive
Free access to archived material in more than 100 Elsevier journals after a specified embargo period.
https://www.elsevier.com/about/open-science/open-access/open-archive.

Elsevier Open Access
Open Access to peer-reviewed articles in journals published by Elsevier that, once accepted into this repository, are free to read and download.
https://www.elsevier.com/about/open-science/open-access/open-access-journals.

Wiley Open Access
All research articles published in Wiley Open Access journals are immediately free to read, download, and share.
http://www.wileyopenaccess.com/details/content/12f25d1df44/About.html.

Data Repositories

Hosted by governments, research institutes, and universities, repositories for large data sets have become more and more common as researchers are being required to deposit their supporting data for a scientific article or research project. Data centers, such as GenBank (genetic sequence data at the US National Center for Biotechnology Information) and PANGAEA (earth and environmental science data hosted by the Alfred Wegener Institute in Germany), house these vast quantities of information. A list of recommended data repositories from *Scientific Data*, a peer-reviewed Open Access journal from Nature Publishing that contains descriptions of scientifically valuable data sets, can be found at http://www.nature.com/sdata/policies/repositories.

Global Biodiversity Information Facility (GBIF)
Funded by governments around the world, GBIF is an international open data source that gives access to data about all types of life on Earth. Its vision is "A world in which biodiversity information is freely and universally available for science, society and a sustainable future." http://www.gbif.org/.

Grey Literature

Zoological information can also be found in grey (or gray) literature, an often hard-to-find type of material that does not appear in traditionally published books and journals. The *Online Dictionary of Library and Information Science* defines grey literature as:

> Material in print and electronic formats, such as reports, preprints, internal documents (memoranda, newsletters, market surveys, etc.), theses and dissertations, conference proceedings, technical specifications and standards, trade literature not readily available through regular channels because it was never commercially published/listed or was not widely distributed.
>
> *Gray literature (2014)*

In the United States, much of the grey literature of science and technology is indexed and made accessible through the National Technical Reports Library (NTRL) database, a collaborative effort of the National Technical Information Service and others, which is available at https://www.ntis.gov/. Grey literature in Europe is included in Open Grey, an Open Access repository containing technical and research reports in science, technology, economics, social science, and humanities, available at http://www.opengrey.eu/.

DISCIPLINE INFORMATION LITERACY INSTRUCTION

Information Literacy Basics

An information literacy session for zoology, or any subject for that matter, should start with the basics on how information is organized and where it will most likely be found. Understanding the concept of the information cycle, that is, "how information comes to be, how it is communicated, and how it should be used appropriately, is designed to illustrate that the whole process of science information affects how one looks for and finds information.

In other words, understanding how science information is produced, shared, and organized is integral to understanding how to find and use it" (Bussman & Plovnick, 2013).

The next important factor is for students to recognize and understand the difference between types of information sources (e.g., popular magazines, scholarly journals, and trade/professional publications). For an in-class library session, one effective way to accomplish this is to set out issues of the different kinds of publications on a table in the room and have the students choose one to examine; this also works well as a group exercise. After showing the table in Fig. 7.1 that describes the types of sources and the key characteristics that differentiate them (purpose, audience, etc.), ask the students to discuss what title they selected, what type of publication they think it is, and give the reasons why (UAA/APU Consortium Library, 2017a).

Another benefit to this exercise is that students are able to see what an actual journal or magazine looks like. For the most part, students today primarily use online articles, so it is valuable for them to see how a journal or magazine issue appears in print—the content, the layout, and any other type of included material that gives context to the publication. With practice, students will begin to compare and contrast the information found in magazines with the scholarly information found in academic, peer-reviewed publications (Carle & Krest, 1998). Students also need to be aware that most online resources allow a search to be limited by publication type or, as it sometimes referred to, document type.

Introducing students to the scientific literature as early as possible is important, since a greater familiarity and awareness of how research is conducted can help them grasp the concepts found in their textbooks and taught in lectures (Porter, 2005). One approach to help students gain such understanding is for them to learn how to recognize scientific articles and how they are organized and presented. This goal can be accomplished by identifying the parts of a scientific article. In general, a typical peer-reviewed scientific paper includes the following sections:

- Title—subject of the paper and what was studied.
- Abstract—summary of the paper, including the reason for the study, the primary results, and the main conclusions.
- Introduction—general background, and sometimes a brief literature review; why the study was done.
- Methods and Materials—how the study was performed, including techniques and equipment, so that it can be replicated by others.
- Results—what happened during the study.

	Popular	Scholarly	Trade / Professional
Purpose	To inform and entertain the general reader	To communicate research and scholarly ideas	To inform readers about a given profession
Audience	General public	Other scholars, students	Practitioners in the field, professionals
Coverage	Broad variety of public interest topics, multiple subjects	Very narrow and specific subjects	Information relevant to a profession
Publishers	Commercial publishers	Professional associations, academic institutions, and commercial publishers	Professional associations or trade groups
Authors	Employees of the publication, freelancers (including journalists and scholars)	Scholars/academics, researchers, experts (usually listed with their institutional affiliation)	Members of the profession, journalists, researchers, scholars
Characteristics	• Little technical language or jargon • Few or no cited references • Absence of list of sources used • General summaries of background information • Extensive ads throughout • Articles are usually brief; about 1–7 pages	• Little or no background information given • Technical language and discipline-specific jargon • Peer review, editorial board • Cites sources throughout • Includes list of sources used • Articles not interrupted with ads • Procedures and materials often described in detail • Articles are longer, often over five pages	• Application of new technology • Employment issues • Practitioners viewpoint • Technical language used • Interpretation of research trends and issues • May include list of sources • Articles are usually brief; about 1–7 pages • Contains ads
Frequency	Frequent, on a daily, weekly, or monthly basis	Less frequent, on a monthly, quarterly, or annual basis	Frequent, on a daily, weekly, or monthly basis
Examples	*National Geographic, Psychology Today, Rolling Stone, Science News, Alaska Dispatch News, Sports Illustrated*	*Journal of American History, Psychological Review, Nature, Contemporary Accounting Research, Journal of Higher Education, American Journal of Sports Medicine*	*American Libraries, Advertising Age, Professional Pilot, Public Manager, Mayo Clinic Health Letter, Chronicle of Higher Education*

UAA/APU Consortium Library, 2017; Modified with permission from Popular Literature vs. Scholarly Peer-Reviewed Literature: What's the Difference? 2017, Rutgers University Libraries.

Figure 7.1 Publication types: a comparative chart. *UAA/APU Consortium Library, 2017; Modified with permission from Popular literature versus scholarly peer-reviewed literature: What's the difference? 2017, Rutgers University Libraries.*

- Discussion—what the results mean and why they are significant.
- Conclusion—summary of the results; may include reasons why further research is needed.
- List of References (or Bibliography)—provides documentation of the sources consulted for the study.

The process continues with an explanation of the difference between primary and secondary, and sometimes tertiary sources. This is a critical piece that students often do not understand, since primary sources in the sciences (e.g., original research articles, proceedings, etc.) are quite different from traditional examples in the humanities and social sciences (e.g., diaries, autobiographies, etc.). Provide copies of articles that are examples of primary and secondary sources for students (or groups of students) to read. Then, ask the students what type of source they think it is and discuss the reasons for their decision. The examples provided should include a few literature reviews so that the students can learn where to find the clues in a given article to help make their decision. Some are obviously reviews and are labeled as such; others are not clearly identified and require more careful reading.

At the end of this portion of the session, ask the students if a limit option is available in the resource or database they plan to search. Most will likely say yes and are quite disappointed to learn that, unless review articles are specified in a search, it is only through a quick (and sometimes not so quick) glance at the article in question that they can determine whether the source is primary or secondary.

The next step is to discuss the peer-review process and its role in scientific publishing. Sometimes the professor has already covered this in the lab or lecture portion of the class, so only a brief review and/or opportunity for questions is necessary. Questions may arise as to the difference between peer review and scholarly, as some resources list both options as limiters. The distinction is somewhat nuanced, but essentially, all peer-reviewed articles are scholarly but not all scholarly articles are peer reviewed. The students also need to know that a peer-review limit option is available but only to the publication level, since not everything published in a peer-reviewed journal is peer reviewed; book reviews, editorials, opinion pieces, and short news items are not considered scholarly articles.

The final step is evaluating sources to determine the quality of the sources the students will come across as they search for information. As stated above, not all items in peer-reviewed journals have gone through the peer-review process, so it is necessary for students to be able to critically evaluate the sources they find. This is especially important when deciding whether or not to include websites as credible sources. While many sites on the Web are authoritative, many more are not. To solve this dilemma, a number of clever and creative acronyms for tools and worksheets that represent the idea of identifying credible sources have been developed. At the University of Alaska Anchorage, librarians who teach information literacy have found the CRAAP Test (see Fig. 7.2) to be a useful and rather lighthearted way to help students remember the good stuff from the "other." CRAAP stands for Currency, Relevance, Authority, Accuracy, Purpose/Point of View, and the test poses a series of short questions to ask whenever the quality of a source is in question (UAA/APU Consortium Library, 2017b).

The various methods just described may seem to be a long drawn-out process to introduce students to zoological information. However, this approach to information literacy can be incorporated in whole or in part—and in tandem with student learning outcomes or objectives. Each can be expanded to allow more time for further discussion and examples on each point in a situation where multiple sessions over the semester are possible, or be done efficiently to allow plenty of time for searching a particular database or resource if time is limited.

Searching for Zoological Information

"In order to be adequately aware of the literature available, it is equally important for students to be aware of the search methods" (Porter, 2005, p. 335). Once students are familiar with the concepts of information and how it is organized, introduce them to Google Scholar or one of the databases previously described in the section on zoology databases. Or, perhaps your library offers access to the literature through a web-scale discovery service such as Ebsco or ProQuest's Summon. Either way, there are some common search methods that universally apply across most databases. These methods include the use of Boolean operators (AND, OR, NOT) and truncation or a wild card character (usually the asterisk) to search a word with multiple endings (e.g., method★ to retrieve method, methods, methodology, etc.). Demonstrating phrase searching by using quotation marks, comparing a basic search to a more complex one by using the advanced search feature, and changing the search fields from keywords to more specific elements such as subject headings or author are additional examples of how searches can be refined. When students see how these tips and techniques improve their search results and make them more relevant, they begin to understand that searching for information is not just a matter of entering some words in a search box and hitting

The CRAAP Test is a list of questions to help you evaluate the information you find. Different criteria will be either more or less important depending on your situation or need.

Currency

How recent is the information?

- Is it current enough for your topic?
- Has it been published in the last x years? (x will vary, depending on your topic.)
- If it is a historical research topic, was it published around the date of the original event?
- Do the links work?
- Has it been updated recently?

Relevance

Does the information address your needs?

- Are references or sources for data or quotations included?
- Where does the information come from and does it apply to your topic?
- Is it a primary or secondary source?

Authority

What is the source of the information?

- Who is the author/publisher/source/sponsor?
- What are their credentials and are they provided?
- What is their reputation or expertise? Are they qualified to write on the topic?
- Is there contact information, such as a publisher or email address?
- What is the domain ending (e.g., .com, .org, .gov, .edu)?
- Are there any advertisements or other distractions?

Accuracy

Is the information reliable, truthful, and correct?

- Is it accurate? Is it supported by evidence?
- Is the information balanced or biased?
- Was it peer reviewed?
- Can you verify the information from another reliable source?
- Are there spelling, grammar, or typographical errors?

Purpose / Point of View

Why does the information exist?

- What is the purpose of the information? Is it to inform/teach/sell/entertain/persuade?
- Is it fact or opinion?
- Who is the intended audience?
- Is this a first-hand account of an event or research?
- If controversial, are all sides of the issue fairly represented?
- Are there political/ideological/cultural/religious/institutional/personal biases?
- Could the site be ironic, like a satire or a spoof?

Figure 7.2 The CRAAP Test (revised). *UAA/APU Consortium Library, 2017; CRAAP Test, revised. Modified and used with permission from Evaluating information: Applying the CRAAP Test, 2010, Meriam Library, California State University, Chico.*

the enter key. Rather, it encourages them to think about what it is that they are really looking for, how to more accurately describe their information need, and to use more sophisticated methods to retrieve the desired results. These are important skills that will make them better information consumers as they progress through their education and beyond.

Undergraduate Students

For undergraduate students, an information literacy session may incorporate some or all of the elements previously discussed. The librarian should be familiar with the assignment or class project, and by using the topics and subject matter as examples for what the students will be researching, the instruction session can be tailored to students' needs. Continuing to emphasize the idea that research is a process of exploration and experimentation will encourage students to overcome their usual tendency to stop the process too soon. Instead, they will begin to dig deeper into their topics and add a wider variety of reference sources and choose more scholarly and peer-reviewed sources (Carle & Krest, 1998).

Graduate Students

A graduate degree in zoology or other scientific discipline is intended to prepare students for jobs in research facilities, laboratories, or the field. Teaching information literacy to graduate students, therefore, builds on the basics discussed earlier but goes on to incorporate more advanced and specialized searching in subject-specific databases. Research topics for graduate students are often quite different from those for undergraduates, since the results will be used to support research projects, grant proposals, and/or graduate theses or dissertations. In general, the smaller number of graduate students in instruction sessions allows the librarian to spend more time with each student, suggesting different resources to try for more in-depth searching. Engaging graduate students in a writing seminar or encouraging them to schedule a one-on-one research consultation with the librarian outside of the classroom can also be effective.

FURTHER THOUGHTS

Ideally, the science librarian or information professional who is teaching information literacy has a scientific background, education, or training, or has had some experience working with scientists in the past. Such qualifications give credibility and are particularly valuable when developing collaborations with science discipline faculty. Understandably, previous knowledge or training in the sciences is not always possible, but librarians "can be effective if they do not have such training by intentionally developing an understanding of the language and terminology of their discipline" (Pritchard, 2010, p. 388). Becoming familiar with the courses offered and the research interests of the academic faculty in the sciences at that institution is also essential.

According to Winterman (2009), "the ideal approach to information literacy education would have the standards taught as part of a process, embedded deeply in the disciplinary content, and would include ongoing communication if not collaboration between librarians and faculty." To encourage faculty to incorporate information literacy into their courses, be proactive and make an effort to contact them to set up an appointment for a meeting. Have them describe what the information needs of the students in the course(s) will be, and offer to work with them to develop appropriate assignments to meet student learning outcomes. The faculty member and the librarian each bring a different experience and expertise to what can be a mutually beneficial and productive partnership, and working together in this way often leads to further collaborations.

Scheduling multiple sessions at key points during the course is ideal so that library instruction can be course integrated or embedded into the student learning process. Of course this is not always possible, given the amount of material that usually needs to be covered, so a one-shot session may be the only option. Bryan and Karshmer (2015) found that the generally positive results of this particular teaching strategy in the research suggests that "while the one-shot may have its limitations ... it is imperative that librarians and collaborators continue to find new ways to use the one-shot within particular disciplines and in ways that highlight specific skills and/or knowledge" (p. 252).

Finally, it is important to assess whether the information literacy goals or student learning outcomes are being met. Multiple methods, limited only by your creativity and the time available, can be employed to assess whether students are understanding the concepts presented. Implementing surveys, either one you have created or one of the many available on the Web, giving pre- and/or postclass quizzes and tests, or using clickers are a few examples of assessment techniques. At the end of a one-shot session, having the students list one thing they learned that they did not know before is another way for the librarian to know whether at least some of the information was successfully presented. For course-integrated sessions given multiple times over a term or semester, getting feedback on a regular basis is essential so that the material presented can be reviewed or clarified in the next library session if students have failed to understand key points. Conversely, presenting more advanced concepts sooner may be possible if the class clearly has a good grasp of the information. Measuring or assessing the results of teaching information literacy as a way to show that students are learning is important, both to demonstrate the value of the library to the institution and, often, to meet requirements for accreditation purposes.

EMERGING TRENDS AND STAYING UP-TO-DATE

The 2016 transition by the Association of College and Research Libraries (ACRL) Board to its *Framework for Information Literacy for Higher Education* has been discussed in previous chapters. Librarians and others in the information profession will no doubt be adjusting their institutions' purpose and mission statements to correspond with the Framework's new directives. However, the basic tenets of information literacy remain the same: the ability to identify, find, and evaluate information across a wide variety of information sources and to use the information ethically and legally.

Articles and other sources on teaching information literacy in the life sciences disciplines appear somewhat regularly in the literature, but tend to be scattered across a variety of publications making them not always easy to find. The peer-reviewed Open Access publication of the Science & Technology Section of ACRL, *Issues in Science & Technology Librarianship*, often has examples of ideas and assignments in various science-related disciplines that can be adapted or modified for your purposes, available at http://www.istl.org/.

ACRL's Instruction Section maintains an Information Literacy in the Disciplines Guide, available at http://acrl.ala.org/IS/is-committees-2/committees-task-forces/il-in-the-disciplines/information-literacy-in-the-disciplines/. It includes a Biology page with links to standards and curricula developed by various organizations involved in higher education.

CONCLUSION

Whether a zoology student plans to obtain a higher degree through graduate studies, teach, or pursue research opportunities with government agencies or technical positions in industry, "the ability to find, read, and interpret the scientific literature is crucial" (Porter, 2005, p. 342). Establishing an effective information literacy program, whether via one-shot sessions, course-integrated classes, or embedded in a science department, can help to meet this goal. The importance of developing collaborative and cooperative relationships between a librarian or information professional and a science faculty member cannot be stressed enough. These efforts are critical to student success and lifelong learning.

REFERENCES

Association of College and Research Libraries [ACRL] Board. (2016). *Framework for information literacy for higher education*. Retrieved from http://www.ala.org/acrl/standards/ilframework.

Bryan, J. E., & Karshmer, E. (2015). Using IL threshold concepts for biology: Bees, butterflies, and beetles. *College & Research Libraries News, 76*(5), 251–255. Retrieved from http://crln.acrl.org/content/76/5/251.short.

Bussman, J. D., & Plovnick, C. E. (2013). Review, revise, and (re)release: Updating an information literacy tutorial to embed a science information life cycle. *Issues in Science and Technology Librarianship, 74*. http://dx.doi.org/10.5062/F4F769JQ.

Carle, D. O., & Krest, M. (1998). Facilitating research between the library and the science writing classroom. *Journal of College Science Teaching, 27*(5), 339–342.

Convention on Biological Diversity. (n.d.). What is taxonomy? Retrieved from https://www.cbd.int/gti/taxonomy.shtml.

Gray literature. In J. M. Reitz (Ed.). (2014). *Online dictionary for library and information science* Retrieved from http://www.abc-clio.com/ODLIS/odlis_about.aspx.

Porter, J. R. (2005). Information literacy in biology education: An example from an advanced cell biology course. *Cell Biology Education, 4*, 335–343. Retrieved from https://www.ncbi.nlm.nih.gov/pmc/articles/PMC1305896/pdf/i1536-7509-4-4-335.pdf.

Pritchard, P. A. (2010). The embedded science librarian: Partner in curriculum design and delivery. *Journal of Library Administration, 50*(4), 373–396. http://dx.doi.org/10.1080/01930821003667054.

Schmidt, D. (2003). Guide to reference and information sources in the zoological sciences. In *Reference sources in science & technology series*. Westport, CT: Libraries Unlimited.

Sheehy, E. P. (1986). *Guide to reference books* (10th ed.). Chicago: American Library Association.

UAA/APU Consortium Library. (2017a). *LibGuide: Scholarly vs. popular* (Modified with permission from Rutgers University Libraries) University of Alaska Anchorage. Retrieved from http://libguides.consortiumlibrary.org/scholarly_vs_popular.

UAA/APU Consortium Library. (2017b). *LibGuide: Evaluating sources: The CRAAP Test* (Revised and modified with permission from Meriam Library, California State University, Chico) University of Alaska Anchorage. Retrieved from http://libguides.consortiumlibrary.org/evaluation.

Winterman, B. (2009). Building better biology undergraduates through information literacy integration. *Issues in Science and Technology Librarianship, 58*. http://dx.doi.org/10.5062/F4736NT6.

Zoology. (2017). In *OxfordDictionaries.com*. Retrieved from https://en.oxforddictionaries.com/definition/zoology.

APPENDIX 1
Additional Resources

Jodee L. Kuden
University of Alaska Anchorage

After reading "How science goes wrong" in The Economist (2013), as further verified in a 2015 Pew Research Center report, it is no wonder that the American public is skeptical of some scientific research, such as the safety of eating genetically modified foods or foods grown with pesticides. *The Economist* article exposes the fact that some scientific research is not reliable when it is not able to be replicated—the very tenet of science—and briefly describes the competitive pressures on researchers to produce and publish, sometimes ignoring "inconvenient" results. The final point of the article is that faulty or poor research results can create a reason to reject or discredit scientific research, leading to doubt about the validity of any research and once again proving the old adage: once burned, twice shy. However, it does solidify yet again the critical importance of scientific literacy and information literacy.

Earlier chapters detail potential resources to incorporate into library instruction sessions to expand students' literacy and knowledge of the life sciences. The purpose here is twofold: one, to list other pertinent resources not previously, or not explicitly, included; and two, to list resources that provide alternative viewpoints on controversial topics. Identifying alternative viewpoints gives students the opportunity to critique their own understanding and knowledge of a specific life science topic. In addition, these different viewpoints can be useful in generating class discussions and debates.

PROPRIETARY DATABASES

Embase
Supplied by Elsevier via paid subscription, this resource is a highly versatile and up-to-date biomedical database, with an emphasis on drug information. It covers the most important international biomedical literature from 1947 to the present.

Pascal
Supplied by EBSCOHost or Ovid via paid subscription, Pascal has been produced by the Institut de l'Information Scientifique et Technique of

the Centre National de la Recherche Scientifique (INIST-CNRS) since 1973. It provides multidisciplinary and multilingual coverage for science, technology, and medicine with special emphasis on European content.

FREE RESOURCES

As noted elsewhere, many national and international agencies and governments have freely available websites that often include a list of their publications.

United Nations Educational, Scientific and Cultural Organization
The United Nations Educational, Scientific and Cultural Organization (UNESCO) main site supplies much needed international or global information and news on natural sciences topics. Information can be found about water, ecology, marine sciences, gender and Indigenous issues, global climate, and biodiversity, to name just a few.
http://www.unesco.org/new/en/natural-sciences/.

Worldwide Nongovernmental Organizations Directory
This directory identifies nongovernmental organizations (NGOs). The UNESCO and USAID (United States Agency for International Development) sites have directories listing NGOs they work with that could be useful starting places to determine the authority of the information found.
http://www.wango.org/resources.aspx?section=ngodir.
UNESCO: http://en.unesco.org/partnerships/non-governmental-organizations.
USAID: https://www.usaid.gov/partnership-opportunities/ngo.

PubMed Central Canada
This site provides access to a stable and permanent online digital archive of full-text, peer-reviewed health and life sciences research publications. It builds on PubMed, the US National Library of Medicine's biomedical and life sciences database, and is a member of the broader PubMed Central (PMC) International (PMCI) network of e-repositories.
http://pubmedcentralcanada.ca/pmcc/.

Europe PubMed Central
A repository giving access to worldwide life sciences articles, books, patents, and clinical guidelines, Europe PMC provides links to relevant records in databases such as Uniprot, European Nucleotide Archive (ENA), Protein Data Bank Europe (PDBE), and BioStudies.
http://europepmc.org/.

Open Access Repositories: International Repositories
This LibGuide published by Arizona State University Library lists websites of freely available repositories from international countries, libraries, and/or agencies.
http://libguides.asu.edu/openaccessresources/international.

Digital Repository, National Science Library Canada
This online repository for the National Research Council (NRC), Canada, includes authored research and collections housed at the National Science Library.
http://dr-dn.cisti-icist.nrc-cnrc.gc.ca/eng/home/ (English-language version).

US Environmental Protection Agency Environmental Topics
US Environmental Protection Agency's (EPA's) website includes information on such topics as water, climate change, and various forms of pollution.
https://www.epa.gov/environmental-topics.

US National Aeronautics and Space Administration Topics: Earth
This US National Aeronautics and Space Administration (NASA) site contains an amazing number of photographs and videos about space, stars, and planets, as well as other topics related to Earth. Data and other available content cover air and atmosphere, global climate, oceans and ice, and more.
https://www.nasa.gov/topics/earth/index.html.

Professional Societies, Associations, Organizations, and Institutes
In the predigital age, the *Encyclopedia of Associations* was the authoritative print resource used to identify professional societies or associations, organizations, and institutes, but now this information is easily found on the Web. Information on the sites may vary; however, many will include up-to-date details about the organizations and list reports, documents, blogs, and other information they have produced.

Land, Ground, and Soil
The topic of land, ground, and soil in life sciences is threaded through the previous subject chapters, but not separated out within them. Land, ground, and soil influence environmental studies, plant life, animal habitat, agriculture, and ecology as well as many other disciplines in the life sciences. Even the chapter about marine and aquatic sciences is related, since ground and nutrient runoff affects marine and aquatic life. Another aspect

of land worth mentioning is land use information. In addition to the resources noted elsewhere, localized land use information may be held by the city, town, municipality, county, parish, borough, state, or provincial governments and may or may not be available in digital format.

ALTERNATIVE RESOURCES

To locate information on opposing sides of a debate, or if in need of other opinions on a controversial topic, the following resources provide an avenue to track down this kind of material.

Proprietary Databases

This listing includes searchable databases with citations, abstracts, and keywords and may or may not contain full-text content.

Alt-Press Watch
Supplied by ProQuest via paid subscription, this resource is an alternative to mainstream media and provides diverse views and distinct voices from independent, grassroots, and small publications.

ATLA Religion
Supplied by EBSCOHost via paid subscription, this database indexes journal articles, book reviews, and collections of essays in all fields of religion, ethics, theology, and philosophy. Coverage goes back to 1949 with retrospective indexing for some journal issues as far back as the 19th century.

Contemporary Women's Issues
Supplied by Gale via paid subscription, this multidisciplinary full-text database brings together relevant content from mainstream periodicals, gray literature, and the alternative press with a focus on the critical issues and events that influence women's lives in more than 190 countries. Contemporary Women's Issues (CWI) compiles often overlooked and hard-to-find newsletters and NGO research reports to which most libraries do not subscribe. It also includes ephemeral literature from leading research institutes and grassroots organizations, which is rarely indexed or cataloged. Records are indexed by 17 categories, including subject, region, article type, and publication type.

Ethnic Newswatch
Supplied by ProQuest via paid subscription, Ethnic Newswatch (ENW) contains newspapers, magazines, journals, and newsletters of the ethnic, minority, and native press. Designed to provide the other side of the story, ENW titles offer additional viewpoints to those of the mainstream press.

The Left Index
Supplied by EBSCOHost via paid subscription, the Left Index includes diverse literature of the left, with an emphasis on political, economic, social, and culturally engaged scholarship inside and outside academia.

Opposing Viewpoints in Context
Supplied by Gale via paid subscription, this resource covers the prominent, contemporary topics that spark debate and controversy. Presented from various viewpoints and listing vetted references, topics are from the humanities, social sciences, and sciences and technology.

Policy File Index
Supplied by ProQuest via paid subscription, public policy issues, global viewpoints, and both sides of the debate are included. EBSCOHost offers a similar product called **Public Affairs Index**, but various descriptions found for it do not mention information for both sides of an issue.

Philosopher's Index
Supplied by EBSCOHost, Ovid, or ProQuest via paid subscription, this index is designed to help researchers easily find publications of interest in all subject areas of philosophy and related disciplines. Serving philosophers worldwide, it contains records from publications dating back to 1902 that originate from 139 countries in 37 languages.

Free Databases
These databases are freely available on the Web and provide citations, abstracts, and sometimes the full-text content.

Congressional Research Service Reports
This searchable site provides access to the full text of many of the Congressional Research Service (CRS) reports that influence US government policies. Written by CRS staff, they are developed for members of the U.S. Congress and are nonpartisan on current subjects.
https://digital.library.unt.edu/explore/collections/CRSR/.

FactCheck.org
Created and maintained by the Annenberg Public Policy Center of the University of Pennsylvania, this site focuses on politics and policies. Many current science topics are discussed by politicians and influenced by governmental policies.
http://www.factcheck.org/.

Snopes.com
One of the oldest and best known fact checking sites on the Web, Snopes started in 1994 and primarily covers urban legends about any subject that are generated from fallacies, misinformation, rumors, strange and unusual happenings, or gossip. In recent years, Snopes has expanded to include many current news stories as well.
http://www.snopes.com/.

Internet Archives

Climate Mirror
This site is a volunteer effort to back up and mirror US federal climate data. Also listed on this site are links to other efforts to archive climate information.
http://climatemirror.org/.

End of Term Web Archive
Hosted by the California Digital Library, this archive captures and saves federal websites from the 2008, 2012, and 2016 administration transitions.
http://eotarchive.cdlib.org/.

Internet Archive: Wayback Machine
An archive containing more than 20 years of web history, this site provides access to websites, free books, software, and more.
https://archive.org/.

Magazines

Typically intended for a general public audience, magazines are a good source for summaries on hot topics as well as information on the latest research in the science disciplines. Articles are written by staff writers, many of whom have a science background. Tables of contents are searchable on the websites of each publication.

Mainstream

Discover, published by Kalmbach Publishing Company since 1980, covers scientific discoveries and more.
http://discovermagazine.com/.

New Scientist, published by Reed Elsevier, contains articles written by staff with a science background. The content has global coverage but is the leading magazine for European science and technological topics.
https://www.newscientist.com/.

Popular Science, published by the Bonnier Corporation Company, is one of the longest running science magazines, dating back to 1872, and covers all areas of science through articles, blogs, and opinion pieces.
http://www.popsci.com/.
Science News, published by the Society for Science and the Public since 1922, contains articles giving overviews of scientific research.
https://www.sciencenews.org/.

Alternative Views
Christian Science Monitor
This independent news organization covers global issues and provides one article per week on spiritual issues. It describes itself as having:

built a reputation in the journalism world over the past century for the integrity, credibility and fair-mindedness of its reporting. It is produced for anyone who cares about the progress of the human endeavor around the world and seeks news reported with compassion, intelligence, and an essentially constructive lens. For many, that caring has religious roots. For many, it does not. The Monitor has always embraced both audiences.

Christian Science Monitor (n.d.)

http://www.csmonitor.com.

High Country News
This newspaper reports on natural science issues important to the western region of the United States "and is the leading source for regional environmental news, analysis and commentary—an essential resource for those who care about this region" (High Country News, 2017).
https://www.hcn.org.

List of Controversial Issues, Wikipedia

Finally, for a list of controversial scientific topics, look no further than this page found in Wikipedia, described as a list of issues that "are constantly being re-edited in a circular manner, or are otherwise the focus of edit warring or article sanctions" (Wikipedia, 2017). Two categories are relevant to the life sciences—environment, and science, biology, and health—and contained in each are all kinds of topics, including current and old "hot topics" available for students to review. This is a useful source to begin a learning journey through the "real" primary science literature.

https://en.wikipedia.org/wiki/Wikipedia:List_of_controversial_issues.

REFERENCES

Christian Science Monitor. (n.d.). *About*. Retrieved from http://www.csmonitor.com/About.
High Country News. (2017). *About*. Retrieved from https://www.hcn.org/about.
How science goes wrong. *The Economist, 409*(8858), (October 19, 2013), 13.
Pew Research Center. (2015). *Public and scientists' view on science and society*. Retrieved from http://www.pewinternet.org/2015/01/29/public-and-scientists-views-on-science-and-society/.
Wikipedia (Eds.). (2017). *Wikipedia: List of controversial issues*. Retrieved from https://en.wikipedia.org/wiki/Wikipedia:List_of_controversial_issues.

INDEX

'*Note*: Page numbers followed by "f" indicate figures, "t" indicate tables.'

A

ABNJ. *See* Areas Beyond National Jurisdiction (ABNJ)
Accreditation requirements, information literacy supports, 5–6
ACRL. *See* Association of College and Research Libraries (ACRL)
AgNIC. *See* Agriculture Network Information Collaboration (AgNIC)
AGRICOLA, 51
Agricultural Research Service (ARS), 52
Agriculture, 47–48, 72
   agricultural information, 48, 52
   agriculture and environmental data archive, 84
   discipline information literacy instruction, 56–58
   discipline resources, 50–54
   emerging trends and staying up-to-date, 58–59
   information literacy instruction for graduate students, 58
   information-seeking behavior, 54–56
   searching as strategic exploration, 48
Agriculture Network Information Collaboration (AgNIC), 51
AGRIS, 51
ALA. *See* American Library Association (ALA)
Alaska North Slope Natural Gas, 98
Alaska Resources Library and Information Services (ARLIS), 97–98
AlgaeBase, 67
Alt-Press Watch, 130
American Library Association (ALA), 1–2, 7
Animal Diversity Web, 112
Animal sciences, 47–48, 115
Annotated bibliography, 57–58
Annual Arctic Ice Atlas, 94
Antarctic
   Antarctic and Southern Oceans Coalition, 99
   Antarctic Research Atlas, 95
   Antarctic Treaty Database, 99
   polar information in, 91
   regions, 96
Applied science, 17, 48
Aquaculture compendium, 69
AQUASTAT, 77
Aquatic commons, 73–74
Aquatic diversity worldwide, 72
Aquatic science, 63–64. *See also* Plant sciences
   aspects of searching, 64–66
   discipline information literacy instruction, 80–83
   discipline resources, 66–80
Aquatic Sciences and Fisheries Abstracts (ASFA), 71
Archival materials, 100–101
Arctic, polar information in, 91–92
Arctic Council, 99–100
Arctic Eight, 90
Arctic Environmental Atlas, 94
Arctic Portal library, 100
Arctic regions, 96
Arctic Science and Technology Information System (ASTIS), 96
Areas Beyond National Jurisdiction (ABNJ), 68
ARLIS. *See* Alaska Resources Library and Information Services (ARLIS)
ARS. *See* Agricultural Research Service (ARS)
ASFA. *See* Aquatic Sciences and Fisheries Abstracts (ASFA)
Assessments, documents from, 98
Association of College and Research Libraries (ACRL), 1–2, 9, 21, 24–25, 66, 87–88, 125
ASTIS. *See* Arctic Science and Technology Information System (ASTIS)
ATLA Religion, 130
Atlases, 94–95
Authority Is Constructed and Contextual concept, 23, 34–35
Awareness services, 107

135

B

Balaenoptera musculus (*B. musculus*), 64–65
   B. musculus intermedia, 64–65
Barents-Euro Arctic Council, 100
Binomial nomenclature, 64–65, 109–110
Biodiversity Heritage Library Portal (BHL Portal), 74
Biofuels, 48
Biological abstracts, 72
Biology, 109, 125
BIOSIS previews, 72, 113
BioStudies database, 128
Books, 54

C

CAB abstracts, 50
CAB direct, 72
Capstone projects, 32
CATs. *See* Classroom Assessment Techniques (CATs)
Christian Science Monitor, 133
Citizen science, 79–80
Classroom Assessment Techniques (CATs), 40
Climate Mirror, 132
Cold regions bibliographies, 96
Commissions, documents from, 98
Communication skills, 12
Congressional Research Service (CRS), 131
Contemporary Women's Issues (CWI), 130
Conversation, scholarship as, 35–36
Coos Bay Georeferenced Bibliography, 65–66
"Core journals," 71
CRAAP. *See* Currency, Relevance, Authority, Accuracy, Purpose/Point of View (CRAAP)
"Credibility and reliability," 56–57
Critical thinking, 12, 28
Currency, Relevance, Authority, Accuracy, Purpose/Point of View (CRAAP), 120–121

D

Data repositories, 116–117
Data sources, 68, 77–79

Databases, 50–51, 66–68, 71–73, 78, 112–115
Designing information literacy instruction, 27
   collaborating with instructors, 31
   designing with objective–activity–assessment approach, 36–42
   information literacy competencies, 33–36
   life sciences
      curriculum, 27–30
      students characteristics in, 30
   opportunities for integrating information literacy, 31–33
   recommendations and practical advice, 42–43
Dictionaries, 95
Digital Age, 7
Digital Repository, 129
Digital revolution, 3, 8
Digitization of polar collections, 105
Directory of Open Access Books (DOAB), 112
Directory of Open Access Journals (DOAJ), 116
Discipline information literacy instruction, 102–105. *See also* Information literacy (IL)
   graduate students, 123
   information literacy goals, 124–125
   information professional, 123–124
   polar information literacy assignments, 103–105
   searching for zoological information, 121–123
   undergraduate students, 123
Discover magazines, 132
DOAB. *See* Directory of Open Access Books (DOAB)
DOAJ. *See* Directory of Open Access Journals (DOAJ)
Dryad, 77

E

e-Science, 18
EBSA. *See* Ecologically and Biologically Significant Areas (EBSA)
EBSCOHost, 131

Ecologically and Biologically Significant Areas (EBSA), 68
Economic Research Service (ERS), 52
Education and Research Archive (ERA), 100
Elective courses, 30
Elsevier Open Access, 116
Elsevier Open Archive, 116
Embase, 127
Embracing transformative, 9–10
Encyclopedia of Associations, 129
Encyclopedia of Life (EOL), 66–67, 111
Encyclopedia of Life Support Systems (EOLSS), 69, 112
Encyclopedias, 68–69, 95, 111–112
Environment Index, 115
Environmental Knowledgebase OnLine. *See* Environment Index
Environmental Protection Agency (EPA), 75
Environmental Sciences and Pollution Management (ESPM), 72
EOL. *See* Encyclopedia of Life (EOL)
EOLSS. *See* Encyclopedia of Life Support Systems (EOLSS)
EPA. *See* Environmental Protection Agency (EPA)
ERA. *See* Education and Research Archive (ERA)
ERS. *See* Economic Research Service (ERS)
ESPM. *See* Environmental Sciences and Pollution Management (ESPM)
Ethnic Newswatch (ENW), 130
European Nucleotide Archive (ENA), 128
Expedition journals and accounts, 93–94
Extension publications and presentations, 53

F

Faculty researchers, 55
FAO. *See* Food and Agriculture Organization (FAO)
Farming, 47
Federal Science Library catalog, 74–75
Feed science. *See* Feeding animals
Feeding animals, 48
Field guides, 69–70
Fish, 72
FishBase, 67
Fisheries, 72
FishStatJ for experts and scientists, 78
Food and Agriculture Organization (FAO), 51, 71, 75
  fisheries and aquaculture department, 78
Food sciences, 47–48
Framework for Information Literacy for Higher Education, 34–36, 56, 88–89
  Authority Is Constructed and Contextual, 34–35
  information creation, 35
  Information Has Value, 35
  research as inquiry, 35
  scholarship as conversation, 35–36
  searching as strategic exploration, 36
Free databases, 131–132
Free resources, 128–130
  proprietary databases, 127–128
"Fugitive literature," 99

G

Gaia antarctic digital repository, 84
GBIF. *See* Global Biodiversity Information Facility (GBIF)
GeoBase, 72
GeoRef, 73
Global Biodiversity Information Facility (GBIF), 78, 117
Global database, 67, 78
Global information repository, 78
Global plants, 68
Global seafarer study of phytoplankton, 80
Glosbe English-Northern Sami Dictionary, 95
Google, 54, 57
  Scholar, 51, 54, 82, 112–113
  searches, 9
Government
  publications, 97–98
  websites, 74–75
Graduate students, 55, 82–83, 123
  information literacy instruction for, 58
Gray literature, 99–100, 117
Grazing systems, 47–48
Green OA, 106
Ground in life science, 129–130
Group project, 57

H

Hands-on approach, 1, 28–29, 31–32
Harvard's Transdisciplinary Research in Energetics and Cancer Center, 17–18
Hearings, documents from, 98
High Country News, 133
High North Research Documents, 96, 100
Higher education, information literacy and, 9–10
Horticulture, 47–48
Hubert Wenger Eskimo Database, 96
Humanities, 18

I

IACUC. *See* Institutional animal care and use committee (IACUC)
IAMSLIC. *See* International Association of Aquatic and Marine Science Libraries and Information Centers (IAMSLIC)
IATTC. *See* Inter-American Tropical Tuna Commission (IATTC)
ICCAT. *See* International Commission for Conservation of Atlantic Tunas (ICCAT)
ICES. *See* International Council for Exploration of the Sea (ICES)
ICILS. *See* International Computer and Information Literacy Study (ICILS)
ICT. *See* Information and Communication Technology (ICT)
IEA. *See* International Association for Evaluation of Educational Achievement (IEA)
IFLA. *See* International Federation of Library Associations (IFLA)
IJC. *See* International Joint Commission (IJC)
IL. *See* Information literacy (IL)
IMRAD format. *See* Introduction, Methods, Results, and Discussion (IMRAD format)
iNaturalist, 79
Indigenous Studies Portal Research Tool, 96
"Information Age," 2
Information and Communication Technology (ICT), 7
Information creation, 22–23, 35
Information Has Value concept, 23, 35
Information literacy (IL), 1–2, 21, 58, 87–90, 117–121, 119f, 122f. *See also* Discipline information literacy instruction; Polar sciences information literacy
  building case for, 3–4
  competencies, 33–36
  higher education and, 9–10
  IL assignments, questions for, 104–105
  improving scientific literacy using, 24–25
  integration, 48
    opportunities for, 31–33
  international support for, 6–9
  life sciences and, 13–14
  primary scientific literature, 12–13
  programmatic approaches for, 41–42
  sciences and, 10–12
  scientific literacy and, 21–23
  standards for science and engineering/technology, 33–34
  supports accreditation requirements, 5–6
  supports lifelong learning, 4–5
Information Literacy Competency Standards for Higher Education, 33–34
Information overload, 3
Information-seeking behavior, 54–56
  agricultural producers, 56
  faculty researchers, 55
  students, 55
Inquiry
  inquiry-based learning approach, 102
  research as, 35
Institut de l'Information Scientifique et Technique of Centre National de la Recherche Scientifique (INIST-CNRS), 127–128
Institutional animal care and use committee (IACUC), 50
Integrated Taxonomic Information System (ITIS), 66, 111
Inter-American Tropical Tuna Commission (IATTC), 75
"Interdisciplinarians," 18
Interdisciplinary approaches, 24

Interdisciplinary research, 10, 17–18, 88
Interdisciplinary teaching teams, 57
Intergovernmental Oceanographic Commission (IOC), 74
Intergovernmental websites, 75–77
International Association for Evaluation of Educational Achievement (IEA), 6
International Association of Aquatic and Marine Science Libraries and Information Centers (IAMSLIC), 73–74, 83, 92
International Commission for Conservation of Atlantic Tunas (ICCAT), 75–76
International Computer and Information Literacy Study (ICILS), 7
International Council for Exploration of the Sea (ICES), 76
International Federation of Library Associations (IFLA), 6
International Field Guides database, 69–70
International Joint Commission (IJC), 76
International Oceanographic Data and Information Exchange (IODE), 68, 73–74
International Pacific Halibut Commission (IPHC), 76
International Polar Years (IPYs), 90, 96
International Repositories, 129
International support for information literacy, 6–9
International treaties and agreements, 98–99
International Union for Conservation of Nature (IUCN), 79, 111
International Whaling Commission (IWC), 76–77
Internet, 18
Internet archives, 132
Introduction, Methods, Results, and Discussion (IMRAD format), 28–29
Inuktitut Living Dictionary, 95
IOC. *See* Intergovernmental Oceanographic Commission (IOC)
IODE. *See* International Oceanographic Data and Information Exchange (IODE)
IPHC. *See* International Pacific Halibut Commission (IPHC)
IPYs. *See* International Polar Years (IPYs)
iSeahorse, 79
ITIS. *See* Integrated Taxonomic Information System (ITIS)
ITU. *See* United Nations International Telecommunications Union (ITU)
IUCN. *See* International Union for Conservation of Nature (IUCN)
IWC. *See* International Whaling Commission (IWC)

J

Journals, 51–52, 70–71, 115

K

King Philip's Class Ordered Family Genus to Speak (KFCOFGS), 65
Knowledge of discipline, 28

L

Laboratory reports, 31–32
Land in life science, 129–130
Landsat Image Mosaic of Antarctica (LIMA), 95
Lapponica, 97
LarvalBase, 67
LC. *See* Library of Congress (LC)
LC Subject Headings (LCSH), 93
Learning, 17
    Learning activities designing, 39
    objectives development, 37–39
Lecture courses, 28
Left Index, 131
Librarians, 125, 3–4, 6, 27, 32–33, 42–43, 66, 87–88, 102
Library catalogs, 93, 93t
Library instruction sessions, 48
Library of Congress (LC), 90–91
Life sciences, 27–30, 113, 116
    courses, 40–41
    information literacy and, 13–14
    laboratory sections, 28–29
    lecture courses, 28
    lower-division courses, 29
    programmatic approaches for information literacy in, 41–42
    required courses *vs.* elective courses, 30
    students characteristics in life sciences, 30
    upper-division courses, 29–30

Lifelong learning, 4–5, 28
LIMA. *See* Landsat Image Mosaic of Antarctica (LIMA)
Livestock production systems, 47–48
Lower-division courses, 29

M

MacKenzie Valley Pipeline Inquiry Report, 98
Magazines, 132–133
  alternative views, 133
  list of controversial issues, Wikipedia, 133
  mainstream, 132–133
Maps, 94–95
  source, 77–79
Marine science, 63–64. *See also* Plant sciences
  aspects of searching, 64–66
  discipline information literacy instruction, 80–83
  discipline resources, 66–80
MarinLit, 73
Massive Open Online Courses (MOOCs), 8–9
Media and Information Literacy (MIL), 7
Media types, 7–8
Messenger, James R., 2
Metaliteracy, 9, 89
MIL. *See* Media and Information Literacy (MIL)
Mobile telecommunication systems, 3
MOOCs. *See* Massive Open Online Courses (MOOCs)
Movie clip, 56–57

N

NAFO. *See* Northwest Atlantic Fisheries Organization (NAFO)
NAL. *See* National Agricultural Library (NAL)
NASS. *See* National Agricultural Statistics Service (NASS)
National Aeronautics and Space Administration (NASA), 129
National Agricultural Library (NAL), 51
National Agricultural Statistics Service (NASS), 52

National Center for Biotechnology Information (NCBI), 67
National Library of Medicine (NLM), 50–51, 113
National Oceanic and Atmospheric Administration (NOAA), 75
National Research Council (NRC), 17, 129
National Science Education Standards, 20
National Science Library Canada, 129
National Technical Reports Library (NTRL), 117
National Weather Service, 53
Natural Resources Conservation Service (NRCS), 53
Natural science collection, 114
NCBI. *See* National Center for Biotechnology Information (NCBI)
New Literacies Alliance, 58
New Scientist magazines, 132
NGO. *See* Non-Governmental Organizations (NGO)
NLM. *See* National Library of Medicine (NLM)
NOAA. *See* National Oceanic and Atmospheric Administration (NOAA)
Non-Governmental Organizations (NGO), 8, 63, 82f, 128
Nordic Cooperation, 95
Nordic Dictionaries, 95
Northwest Atlantic Fisheries Organization (NAFO), 77
NRC. *See* National Research Council (NRC)
NRCS. *See* Natural Resources Conservation Service (NRCS)
NTRL. *See* National Technical Reports Library (NTRL)
Nunavik Marine Region Wildlife Board, 98
Nunavut Arctic College, 101

O

OA. *See* Open Access (OA)
OBIS. *See* Ocean Biogeographic Information System (OBIS)
Objective–activity–assessment approach, 36–42, 36f

designing learning activities, 39
learning objectives development, 37–39, 38f
programmatic approaches for information literacy, 41–42
student learning assessment, 40–41
Ocean Biogeographic Information System (OBIS), 68
OceanDocs, 74
Oceanic abstracts, 73
Oceanography, 63–64
OECD. *See* Organization for Economic Cooperation and Development (OECD)
Oncorhynchus tshawytscha (O. tshawytscha), 64
Online Dictionary of Library and Information Science, 117
Open Access (OA), 92, 106
   gateway, 7
   journals, 70–71, 115–116
   repositories, 129
   scholarly, 106
Oregon Estuarine Invertebrates, 84
Organization for Economic Cooperation and Development (OECD), 20

P

Pacific Salmon Commission (PSC), 77
Pan Inuit Trails Atlas, 94
PANGAEA, 78
Panthera tigris altaica, 110
Pascal, 127–128
Peer-review process, 120
Philosopher's Index, 131
PISA. *See* Programme for International Student Assessment (PISA)
Planning information literacy instruction, 29–30, 43, 56
Plant sciences, 47. *See also* Aquatic science; Marine science
   discipline information literacy instruction, 56–58
   discipline resources, 50–54
   emerging trends and staying up-to-date, 58–59
   information literacy instruction for graduate students, 58
   information-seeking behavior, 54–56
   searching as strategic exploration, 48
PLC. *See* Polar Libraries Colloquy (PLC)
PLoS. *See* Public Library of Science (PLoS)
PMC. *See* PubMed Central (PMC)
PMC International (PMCI), 128
PNRA. *See* Programma Nazionale Di Ricerche in Antartide (PNRA)
Polar, 89
   availability of polar materials in languages, 106
   digitization of polar collections, 105
Polar Code MECP 68-21 Annex 10, 99
Polar information, 90–92
   Antarctic, 91
   archival materials, including photos and videos, 100–101
   Arctic, 91–92
   documents from hearings, commissions, assessments, 98
   expedition journals and accounts, 93–94
   government publications, 97–98
   gray literature, 99–100
   international treaties and agreements, 98–99
   library catalogs, 93, 93t
   literacy assignments, 103–105
   polar data repositories, 101
   polar databases, 96–97
   polar journals, 95–96
   polar libraries, 92
   reference tools, 94–95
   sources, 90
   TK, 101–102
Polar Libraries Colloquy (PLC), 92
Polar Mapping Sites, 94–95
Polar sciences information literacy, 87. *See also* Information literacy (IL)
   discipline information literacy instruction, 102–105
   emerging trends and staying up-to-date, 105–107
   information literacy in polar context, 87–90
   listing of polar information resources, 92–102
   polar information, 90–92

Policy File Index, 131
Popular Science magazines, 133
Primary scientific literature, 12–13
Problem-based learning, 24, 39
Professional societies, associations, organizations, and institutes, 129
Programma Nazionale Di Ricerche in Antartide (PNRA), 91
Programmatic approaches, 27, 41–42
Programme for International Student Assessment (PISA), 20
Project Information Literacy, 4
Proprietary databases
  alternative resources, 130–131
  free resources, 127–128
Protein Data Bank Europe (PDBE), 128
PSC. See Pacific Salmon Commission (PSC)
PubAg, 51
Public Affairs Index, 131
Public communication of science. See Scientific literacy
Public Library of Science (PLoS), 115
PubMed, 50–51, 113
PubMed Central (PMC), 116, 128

R

Reef Life Survey (RLS), 80
ReefBase, 78
Reference books, 111–112
Reference tools, 94–95
  Atlases, Maps, and Polar Mapping Sites, 94–95
  dictionaries, 95
  encyclopedias, 95
Reports, Reviews, Meetings (RRM), 72, 113
Repositories, 73–74
Required courses, 30
Research as inquiry, 35
Research papers, 32
Resources for life science, 127
  alternative resources, 130–133
  free resources, 128–130
  proprietary databases, 127–128
Revised Bloom's Taxonomy, 37f
RLS. See Reef Life Survey (RLS)
RRM. See Reports, Reviews, Meetings (RRM)

Rubrics, 40–41
Russian Web of Science, 97

S

Scholarly literature, 35
Scholarly OA, 106
Scholarship as conversation, 35–36
Science, 22
  communication. See Scientific literacy
  curriculum, 28
  disciplines, 17
  information literacy and, 10–12
  News magazines, 133
Science and Technology Section (STS), 10
Scientific literacy, 19–21
  improving scientific literacy using IL skills, 24–25
  information literacy and, 21–23
Scientific research, 127
Scott Polar Research Institute Library (SPRILIB), 97
Scott Polar Research Library (SPRI), 92
  Polar Directory, 92
Sea around Us, 78–79
SeaLifeBase, 68
Search strategies, 64
Searching as strategic exploration, 36
Secchi disk, 80
Secretariat of the Pacific Community (SPC), 84
Self-reflection, 40–41
Smith-Lever Act (1914), 53
Snopes.com, 132
Social media, 9
Social structure of science, 19
"Soft" skills, 18–19
Soil, 47
  in life science, 129–130
SPC. See Secretariat of the Pacific Community (SPC)
Species databases, 66–68
Spitsbergen Treaty, 99
SPRI. See Scott Polar Research Library (SPRI)
SPRILIB. See Scott Polar Research Institute Library (SPRILIB)
"Stand-up Boolean," 81

Staying up-to-date, 58–59, 83–84
Strategic exploration, 22–23, 36
Strengths, weaknesses, opportunities, threats (SWOT), 24
STS. *See* Science and Technology Section (STS)
Students, 55
 characteristics in life sciences, 30
 learning assessment, 40–41
Summative assessment, 40–41
SWOT. *See* Strengths, weaknesses, opportunities, threats (SWOT)

T

TAGA. *See* Toolik-Arctic Geobotanical Atlas (TAGA)
Task Force, 11
Taxonomy
 databases, 66–68
 taxonomic information, 111
 taxonomy/scientific nomenclature, 109–110
 tools, 65
Teaching Tripod Approach, 37
Tiger, taxonomic classification of, 110
TK. *See* Traditional knowledge (TK)
Toolik-Arctic Geobotanical Atlas (TAGA), 94
Trade publications, 54
Traditional knowledge (TK), 101–102

U

UN Environment Programme/GRID-Arendal, 94
Undergraduate research experiences, 32–33
Undergraduate students, 80–82, 123
UNEP. *See* United Nations Environment Programme (UNEP)
UNEP World Conservation Monitoring Centre (UNEP-WCMC), 79
Uniprot database, 128
United Nations Convention on Law of Sea, 99
United Nations Educational, Scientific, and Cultural Organization (UNESCO), 7, 112, 128
United Nations Environment Programme (UNEP), 79
United Nations International Telecommunications Union (ITU), 3
United States Agency for International Development (USAID), 128
University of Alaska Fairbanks, 94
University of Alberta Circumpolar Collection, 100
Upper-division courses, 29–30
US Department of Agriculture (USDA), 51
US Environmental Protection Agency (EPA), 129
US Geological Survey (USGS), 75, 95
Video clip, 56–57

W

Water Resources of United States, 79
Wayback Machine, 132
Weather data, 53
Web Archive, 100, 132
Web of Science, 114
 core collection, 50
Websites, 52–53
Wikipedia, 57, 133
Wildlife & Ecology Studies Worldwide, 114–115
Wiley Open Access, 116
World Database of Protected Areas (WDPA), 79
World Wide Web (WWW), 3
Worldwide nongovernmental organizations director, 128
 discipline information literacy instruction, 117–123
 discipline resources, 110–117
 emerging trends and staying up-to-date, 125
 searching for zoological information, 121–123
 taxonomy/scientific nomenclature, 109–110
 zoological record, 73, 114